Methods and Applications
of Interval Analysis

SIAM Studies in Applied Mathematics

WILLIAM F. AMES, Managing Editor

This series of monographs focuses on mathematics and its applications to problems of current concern to industry, government, and society. These monographs will be of interest to applied mathematicians, numerical analysts, statisticians, engineers, and scientists who have an active need to learn useful methodology for problem solving.

The first title in this series is *Lie-Bäcklund Transformations in Applications*, by Robert L. Anderson and Nail H. Ibragimov.

Ramon E. Moore

METHODS AND APPLICATIONS
OF INTERVAL ANALYSIS

siam *Philadelphia/1979*

Library of Congress Catalog Card Number: 79-67191
ISBN: 0-89871-161-4

"A Bibliography on Interval-Mathematics," by Fritz Bierbaum and Klaus-Peter Schwiertz, reprinted by permission from *Journal of Computational and Applied Mathematics*, 4 (1978), pp.59-86.

to Adena with love

Contents

Preface

The demands of the computer age with its "finite" arithmetic dictate the need for a structure which has come to be called *interval analysis* or later *interval mathematics*. Since its introduction some twenty years ago the subject has undergone rapid development. It is serving as an impetus for a quest for rigor in *numerical computations on machines*. At the same time it has kindled an interest on the part of mathematicians in developing the foundations of numerical computation.

Anyone using a computer, whether in physical sciences, economic modeling, engineering design or whatever has surely inquired about the effect of rounding error and propagated error due to uncertain initial data or uncertain values of parameters in mathematical models. A standard question should be "what is the error in my result?" This book supplies techniques for keeping track of errors. Algorithms are developed, and applied, for the machine computation of rigorous error bounds on approximate solutions. Computational tests for machine convergence of iterative methods, existence and non-existence of solutions for a variety of equations are obtained via *interval analysis*.

Interval analysis is a new and growing branch of *applied mathematics*. It is an approach to computing that treats an interval as a new kind of number. Computations in properly rounded interval arithmetic produce results which contain both ordinary machine arithmetic results and also infinite precision arithmetic results. Thus, we have at the outset, a completely general mechanism for bounding the accumulation of roundoff error in any machine computation. If roundoff is the only error present, then the widths of the interval results will go to zero as the length of the machine word increases.

An interval has a dual nature as both a *number* and a *set* of "real" numbers. Many of the algorithms for interval methods make use of this duality and combine set theoretic operations such as set intersection with arithmetic operations. In a single interval computation, for a problem in which we allow various coefficients and initial data to range independently over intervals of values, an interval result will contain the entire set of possible values of the solution. The ability to compute with sets of values in interval arithmetic provides for some simple computational tests for existence, uniqueness, and convergence. By operating with interval-valued functions and by extending such concepts as integration and operators to interval-valued functions we can construct upper and lower bounding functions to the solutions of operator equations, such as boundary value problems.

Somewhat independently, efficient methods have been found for the recursive generation of Taylor series coefficients. By combining interval methods with Taylor series methods, we can compute upper and lower bounds to solutions of differential equations even when the differential equations have uncertain coefficients, which, along with the initial data, are known only to lie in certain intervals.

Many of the methods for finding upper and lower bounds on solutions to operator equations based on different inequalities will be seen to be special cases of interval methods (see § 8.4).

The book is intended as a text and reference work for scientists and engineers who, as sophisticated practitioners of the computing art, wish to know about the errors in their results especially in those cases when data is uncertain or poorly described. On the other hand there is material of interest to the applied mathematician interested in new developments in and applications to algebra, recursive function theory, set theory, differential and integral equations and constructive analysis.

The material in this book can be and has been used for a one semester senior–graduate level course. It is accessible to a person with a standard undergraduate training in the mathematical sciences with some knowledge of computer programming.

While the book is one in applied mathematics its orientation is definitely toward computation with special attention paid to physical, engineering and economic mathematical models. A special feature is the inclusion of *A Bibliography on Interval-Mathematics,* by Fritz Bierbaum and Klaus-Peter Schwiertz of Germany.

Since the appearance (1966) of the monograph *Interval Analysis* [57] (now out of print) and its German translation (1969), the subject has undergone remarkable development. The present monograph will help bring the reader up to date on many of the developments of the subject which have appeared during the past decade. It will serve as a guide to the current literature for those who wish more details on a particular application than can be included in a brief treatment.

I am grateful to William F. Ames for inviting me to write this book and for his helpful suggestions for revisions. I am indebted to Adena Spohn Moore for helping me write chapter nine. I have benefited from discussions with many colleagues, students, and friends—above all, Karl Nickel. Some of the research on nonlinear systems was supported by the National Science Foundation under Grants MCS76-83883 and MCS78-03824.

R. E. MOORE
University of Wisconsin
Madison, 1978

Methods and Applications
of Interval Analysis

Chapter 1

Introduction

It is the purpose of this monograph to provide an account of some of the principal methods and applications of *interval analysis*—a new, growing, and fruitful branch of applied mathematics.

The basis for the new methods is an extension of the concept of a "real" number. In Chapter 2 we show that an *interval* of "real" numbers can be thought of as a *new kind of number*, represented by a pair of "real" numbers, namely its endpoints. We introduce an arithmetic—*interval arithmetic*—for such numbers. When the endpoints of an interval are the same, we have a "degenerate" interval—which we can identify with the "real" number equal to its identical endpoints. Thus, interval arithmetic includes "real" arithmetic as a special case.

In practice, we cannot carry out "real" arithmetic anyway. We are confined to approximate arithmetic of limited precision. We commit roundoff errors. By properly rounding the endpoints of the machine-computed results of interval arithmetic (as described in § 2.4), we can compute intervals *containing results of infinite precision*; in other words we can find *intervals* containing the *exact* "real" arithmetic results. We can compute *arbitrarily narrow intervals* containing exact real arithmetic results by carrying enough digits.

Even if we could solve a mathematical equation exactly, which we usually cannot, we would still have only an approximate description of the behavior of a real physical system which the mathematical equation is supposed to model. For one thing, the constants in the mathematical model will usually be known only approximately. In particular, if there are constants which are experimentally determined by measurements, then there will be limited precision in the numerical values assigned to the quantities. With interval methods, as presented in this monograph, we will be able to compute bounds on the *set* of solutions corresponding to intervals of possible values for the measured quantities. "Although interval analysis is in a sense just a new language for inequalities, it is a very powerful language and is one that has direct

1

applicability to the important problem of significance in large computations",
(R. D. Richtmeyer, Math. Comput., 22 (1968), p. 221).

At the end of each chapter will be found notes pointing to specific references
and occasional comments. A notation such as (N 1) in the text refers to note 1
at the end of that chapter. Proceedings will be listed only once, with the name of
the editor. Individual papers will be referred to by page numbers. At least
seven hundred works have been written to date in some form or other on
interval mathematics (N 5). However, only a fraction of these are accessible
in readily available books or journals. Many are in the form of internal reports
of research institutes, doctoral dissertations, etc. and are not readily available.
In this monograph, references are confined, as much as possible, to the open
literature. Much of the literature is European due to the work and influence of
two of the early innovators in the field, F. Krückeberg and K. Nickel, and their
colleagues and students.

Applications of interval methods which have been reported to date include
such diverse areas as mathematical programming (Chapter 7), operator equations (Chapter 8), geodetic computations (N 1), analysis of electrical circuits
(N 2), mathematical psychology (N 3), and the re-entry of an Apollo-type
spaceship into the earth's atmosphere (N 4). Actually, the methods have
application to any computation whatever, at the very least to bound accumu-
lated roundoff error and propagated error in initial data.

Perhaps *uncertainty* in initial data would be a more accurate description for
many applications than "error" in initial data. Much of the highly repetitive
computing in "production" runs carried out in applied mathematics involves
case studies of the behavior of mathematical models: parameter studies,
sensitivity analysis, design studies, effects of inaccurate measurements or
observational errors, etc. We carry out dozens, hundreds, or even thousands of
runs with different numerical values for initial data and parameters in the
mathematical model. In many such situations, we could compute the quantities
of interest with a *single* interval computation, leading to a *set* of possible values
corresponding to allowable ranges for the parameters and initial data. As an
example, suppose we wish to know how much a particular real root of a
polynomial can change when we vary the coefficients by certain amounts. Say
we have found an approximate root r^* for nominal values of the coefficients. If
we can find r_1 and r_2 such that $r_1 < r^* < r_2$ and such that $p(r_1, C) < 0$ and $p(r_2, C) > 0$, say, where $p(r_1, C)$ and $p(r_2, C)$ are intervals obtained from evaluating
the polynomial in interval arithmetic (§ 2.2) with coefficients lying in an interval
vector C (§ 2.1), then we will have found that there is at least one root r^* in the
interval $[r_1, r_2]$ when the coefficients lie in C. Furthermore, a *single* interval
arithmetic evaluation of the polynomial can yield an interval containing the
range of values $p(R, C)$ corresponding to any r in an interval R and any set of
coefficients in C. If the resulting interval does not contain zero, then there are
no roots in R for *any* such coefficients. Such ideas can be, and have been, used

to compute upper and lower bounds on roots of polynomials. For instance, in performing a cost-benefit analysis of a project, the World Bank measures the project's profitability by computing the rate-of-return of a projected cost-benefit stream. The rate-of-return is a root of a polynomial whose coefficients are the projected cost-benefit stream. Interval methods have been used by the World Bank to find upper and lower bounds on rates-of-return (N 6). See Chapter 9.

Applications of interval methods to finding upper and lower bounds to solutions of operator equations (differential equations, integral equations, etc.) are discussed in Chapter 8. In § 8.4 it is shown that certain interval methods are generalizations of methods based on differential inequalities and monotone operators. On the other hand, there are computational methods which *combine* interval techniques with those derived from the theory of differential inequalities (§ 8.4).

We are *not* suggesting that *all* computation should be carried out using interval techniques, but only that interval methods provide another set of *tools* in applied mathematics—tools which can be especially helpful in analyzing computational error, verifying sufficient conditions for existence and convergence, constructing upper and lower bounds on sets of solutions, and in providing natural stopping criteria for iterative methods.

We can represent certain regions in n-dimensional Euclidean space, specifically n-dimensional rectangles, by interval vectors, that is n-tuples of intervals $(X_1, X_2, \cdots, X_n)$. By carrying out a finite sequence of interval arithmetic operations beginning with intervals $X_1, X_2, \cdots, X_n$ we obtain an interval containing the range of values of the corresponding real *rational function* when the real arguments $x_1, x_2, \cdots, x_n$ vary independently over the intervals $X_1, X_2, \cdots, X_n$.

We can find interval extensions of *nonrational functions* (functions which are not rational), including all the elementary functions: exponentials, logarithms, trigonometric functions, etc. The class of functions expressible as rational combination and compositions of arithmetic and elementary functions contains most of the functions used in computation and in applied mathematics (see § 4.4). A variety of techniques is available for refining upper and lower bounds on ranges of values of functions (§ 4.4). We can, in various ways, compute *arbitrarily sharp* upper and lower bounds on ranges of values.

By using interval arithmetic, interval-valued functions, integrals of interval-valued functions (§ 4.5), etc., we can compute arbitrarily sharp upper and lower bounds on exact solutions of many problems in applied mathematics. Current research centers around studies of the *most efficient* ways to do this for particular classes of problems.

Some of the interval algorithms are extensions of corresponding real algorithms. Some are essentially different. For example, many interval algorithms involve taking *intersections* of two or more intervals. There is no corresponding operation for real numbers. An interval number (Chapter 2) is also a *set*. As a result, we can and do combine arithmetic with set theoretic operations in interval algorithms.

Computationally verifiable sufficient conditions for the *existence* or *non-existence* of a solution to a given problem in applied mathematics can be formulated using interval methods (Chapter 5). At the same time, computationally verifiable sufficient conditions for the *convergence* of iterative algorithms can be given using interval methods (Chapter 5).

An interesting and useful aspect of interval algorithms is that of *finite convergence* (Chapter 4). If an algorithm produces a nested sequence of intervals using exact interval computation, then it will converge in a finite number of steps (to some interval) when the computations are carried out in rounded interval arithmetic if each new iterate is intersected with the previous one. In finite precision *rounded* interval computation such a procedure will always reach a point beyond which all iterates are exactly equal. This gives us a convenient stopping criterion for such interval algorithms (§ 4.3): we stop when two iterates are exactly equal.

The two main *objections*, in the past, to the use of interval methods have been: (1) the resulting bounds may be excessively pessimistic and (2) the interval computations take too much extra time. We will discuss each of these objections in turn.

It is one of the goals of this monograph to show that *arbitrarily sharp* bounds on exact solutions for various types of equations can be obtained with interval methods. Suppose, for example, that we can carry out, in rounded interval arithmetic, the solution of a linear algebraic system using Gaussian elimination. We will assume in this illustration, as is usually assumed, that the coefficients of the linear system can be exactly represented by machine numbers. Suppose we obtain an interval vector containing the solution vector (which exists, by the way, if the interval procedure can be carried out!—see § 5.1), but that the components of the resulting interval vector are wide—say having widths about 1.0. If we carry out the same computation again, but carry the equivalent of 8 more decimal digits (say "double precision" instead of "single precision"), we may well obtain an interval vector (containing the solution) whose components have widths about 10^{-8}. (See § 5.1 for an illustration of such an effect.) By various techniques to be discussed in this monograph we can obtain arbitrarily sharp bounds on exact solutions and even arbitrarily sharp bounds on *sets* of solutions (corresponding to intervals of coefficients) to problems in applied mathematics. In other words, there is no reason to accept interval results which are not sufficiently narrow in width when we can improve the bounds by further computation, except, possibly, the cost of such further computations. In short, the first objection really can be refuted if we are willing to do enough computation.

This brings us to the second objection, which is really the only one of the two that can be taken seriously. If rounded interval arithmetic is programmed on a computer using subroutine or procedure calls in a "high level language" such as FORTRAN or Algol, then a rounded interval arithmetic operation will require much more CPU time for its execution than the corresponding hardware machine arithmetic operation. On the other hand, computing machines being built today have "writeable control stores" so that rounded interval arithmetic can be put into the hardware, so to speak. We can *microprogram* rounded interval arithmetic operations in the writeable control store so that they can then be executed at speeds comparable to those of ordinary "hardware" machine arithmetic. When this becomes common practice, the second objection will *finally* disappear, too! We will then be able to compute bounds on the range of values of a function over an n-dimensional rectangle in about the same CPU time required to obtain a single (approximate) value of the function. In the meanwhile, the second objection may be a serious one for large problems requiring extensive computation. For many problems, however, it will usually be of no practical consequence whatever whether an important computation requires a fraction of a second or several minutes for its completion on a computer. The total cost will still be small. In such a case, interval methods can provide a *rigor* in the computations (verification of the existence of a solution and guaranteed, rigorous error bounds on roundoff error, propagated initial error, truncation error, etc.) *which seems unattainable any other way.* It will often be worth the small extra cost. A major goal of current research in interval analysis is that of finding *efficient* interval methods (see, for example, § 4.4). We would like to find methods which are as fast as possible for computing rigorous error bounds on approximate solutions. Further progress in this area can be expected for some time to come.

If we can compute an interval $[a, b]$ *containing* an exact solution x to some problem, then we can take the *midpoint, $m = (a + b)/2$,* of the interval as an approximation to x and we will have $|x - m| \leq w/2$, where $w = b - a$ is the *width* of the interval $[a, b]$. Thus, the computation of an interval containing an exact solution provides at once both an approximation to the solution and error bounds on the approximate solution.

In § 2.3 we discuss algebraic properties of interval arithmetic. It will be seen, in § 2.3 and in § 3.2, that fundamental relations exist between the algebraic and the set theoretic properties of interval numbers.

In Chapter 3 we discuss the finite evaluation of set valued mappings. Interval valued functions are defined which are extensions of rational as well as irrational real valued functions. A single evaluation of an interval extension provides upper and lower bounds on the range of values of a real valued function whose arguments may vary independently over intervals of possible values.

In Chapter 4 it is shown how we may compute arbitrarily sharp upper and lower bounds on the range of values of a function over an n-dimensional rectangle; in § 4.4 we investigate the question of the most *efficient* ways to bound ranges of values.

Continuity for interval functions is defined, in § 4.1, in such a way that it is a natural extension of continuity for real valued functions.

A general, set theoretic property of arbitrary mappings (the *subset* property) is explained in § 3.2. It is seen (Chapters 3 and 4) that natural interval extensions enjoy the related property which we call *inclusion monotonicity*, not to be confused with monotonicity in the sense of an increasing real valued function (which is to say, monotonicity with respect to the order relation ≦). Inclusion monotonicity is defined with respect to the set-inclusion partial order relation ⊆ ; the underlying "subset property" simply says that, for an arbitrary mapping *f*, if *A* and *B* are subsets of the domain of *f* and if *A* is a subset of *B*, then the image of the subset *A* under the mapping *f* is contained in the image of *B*. Many of the methods of interval analysis make use of this important property.

"It used to be part of the folklore of numerical analysis that . . . derivatives are difficult to evaluate . . ." (P. Henrici; see Chapter 3, (N 5)). We show, in § 3.4, that derivatives are *not* difficult to evaluate if we do it *recursively* and if we store a certain array of intermediate results. While the method would be cumbersome for hand computation, it is *easy* on a stored program computer. As a result, we can carry out successive Taylor series expansions efficiently for a variety of purposes (quadrature, solution of initial value problems, etc.) with remainder (in mean value form) computed using interval methods.

In § 3.5 we discuss the *enclosure* of irrational functions in polynomials with interval coefficients. A technique is given for reducing the degree of the resulting interval polynomial enclosure.

The concept of a *Lipschitz* interval extension is introduced in § 4.1. This is useful in later sections in tests for the existence of solutions to operator equations. Many of the interval extensions we will use in computing have the property.

In § 4.5 we discuss the integration of interval valued functions of a real variable, as well as interval methods for computing upper and lower bounds on exact values of integrals of both real and interval valued functions. Formal integration of interval polynomials is also defined.

In Chapter 5 we present computable tests for existence of solutions and convergence of iterative methods for linear systems, nonlinear systems, and operator equations in function spaces. Section 5.2 contains recent results for finite systems of nonlinear equations.

An interval version of Newton's method for a nonlinear equation in a single variable is seen, in § 5.2, to possess especially remarkable convergence properties, particularly when carried out in *extended* interval arithmetic (in which we allow unbounded intervals). This, too, can be carried out on a computer.

Operator equations (differential equations, integral equations, etc.) are considered in § 5.3 and in Chapter 8. We present interval methods for verifying the existence of solutions and for constructing upper and lower bounding functions.

In Chapter 6, we discuss a problem which is just beginning to receive some attention, in spite of its long standing importance. Namely, the problem of *finding* a safe starting point for an iterative method for solving a nonlinear system of equations. We present an approach, based on interval computation, using *exclusion tests* (nonexistence of a solution in a test region). An *n*-dimensional *interval bisection* procedure is given, whose *stopping criterion* is the satisfaction of computationally verifiable tests for existence of a solution in a test region and convergence of a given iterative method from the test region (or from an arbitrary point in the test region). This is a new area and research is still in progress, looking for improved methods.

In Chapter 7, some applications of interval methods to mathematical programming are discussed. Useful applications have been made in *linear programming* (§ 5.1), *convex programming* (§ 7.2), and general (nonconvex), differentiable, nonlinear optimization.

The notes at the end of chapters and the references at the end of the monograph will provide a useful guide to the current literature.

NOTES

1. See F. Hussain [35] and G. Schmitt [86].

2. See S. Skelboe [89].

3. See H. Ratschek [70, pp. 48–74].

4. See U. Marcowitz [53].

5. See F. Bierbaum and K.-P. Schwierz [8a].

6. P. Gutterman, Computing Activities Department, The World Bank, Washington, DC, personal communication.

Chapter 2

Finite Representations

2.1. Interval numbers, vectors, and matrices. By an *interval* we mean a closed bounded set of "real" numbers

$$[a, b] = \{x : a \le x \le b\}.$$

We can also regard an interval as a *number* represented by the ordered pair of its endpoints a and b; just as we represent a rational number, a/b, by an ordered pair of integers and a complex number, $x + iy$, by an ordered pair of real numbers. We will presently introduce arithmetic operations for interval numbers. Thus, intervals have a *dual nature*, as we shall see, representing a *set* of real numbers by a new kind of *number*.

We will denote intervals by capital letters. Furthermore, if X is an interval, we will denote its endpoints by $\underline{X}$ and $\bar{X}$. Thus, $X = [\underline{X}, \bar{X}]$.

By an *n-dimensional interval vector*, we mean an ordered n-tuple of intervals $(X_1, X_2, \cdots, X_n)$. We will also denote interval vectors by capital letters. Thus, if X is a two-dimensional interval vector, then $X = (X_1, X_2)$ for some intervals $X_1 = [\underline{X}_1, \bar{X}_1]$ and $X_2 = [\underline{X}_2, \bar{X}_2]$. A two-dimensional interval vector also represents a two-dimensional *rectangle* of points (x_1, x_2) such that $\underline{X}_1 \le x_1 \le \bar{X}_1$ and $\underline{X}_2 \le x_2 \le \bar{X}_2$. See Figure 2.1. We will not distinguish between the degenerate interval $[a, a]$, and the real number, a.

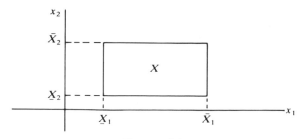

FIGURE 2.1

9

If the real number x is in the interval X, we will write $x \in X$. Similarly, if $x = (x_1, x_2, \cdots, x_n)$ is a real vector and $X = (X_1, X_2, \cdots, X_n)$ is an interval vector, we will write $x \in X$ if $x_i \in X_i$ for $i = 1, 2, \cdots, n$.

We call two intervals *equal* if their corresponding endpoints are equal. Thus, $X = Y$ if $\underline{X} = \underline{Y}$ and $\bar{X} = \bar{Y}$.

The *intersection* of two intervals X and Y is *empty*, $X \cap Y = \varnothing$, if either $\underline{X} > \bar{Y}$ or $\underline{Y} > \bar{X}$. Otherwise, the intersection of X and Y is again an interval

(2.1) $$X \cap Y = [\max (\underline{X}, \underline{Y}), \min (\bar{X}, \bar{Y})].$$

The intersection of two interval vectors is empty if the intersection of any of their corresponding components is empty, otherwise, for $X = (X_1, X_2, \cdots, X_n)$ and $Y = (Y_1, Y_2, \cdots, Y_n)$ we have $X \cap Y = (X_1 \cap Y_1, \cdots, X_n \cap Y_n)$, which is again an interval vector.

If two intervals X and Y have nonempty intersection, their *union*, $X \cup Y = [\min (\underline{X}, \underline{Y}), \max (\bar{X}, \bar{Y})]$, is again an interval. The union of two intersecting interval vectors is *not*, in general, an interval vector.

We can extend the transitive order relation, $<$, on the real line to intervals as follows:

(2.2) $$X < Y \quad \text{if and only if} \quad \bar{X} < \underline{Y}.$$

Another, extremely useful, transitive order relation for intervals is *set inclusion*:

(2.3) $$X \subseteq Y \quad \text{if and only if} \quad \underline{Y} \le \underline{X} \quad \text{and} \quad \bar{X} \le \bar{Y}.$$

If $X = (X_1, \cdots, X_n)$ and $Y = (Y_1, \cdots, Y_n)$ are interval vectors, we have $X \subseteq Y$ if $X_i \subseteq Y_i$ for $i = 1, 2, \cdots, n$.

We denote the *width* of an interval $X = [\underline{X}, \bar{X}]$ by

(2.4) $$w(X) = \bar{X} - \underline{X}.$$

Similarly, the width of an interval vector $X = (X_1, \cdots, X_n)$ is $w(X) = \max (w(X_1), \cdots, w(X_n))$.

We define the *absolute value* of an interval X by

(2.5) $$|X| = \max (|\underline{X}|, |\bar{X}|).$$

Thus, $|x| \le |X|$ for every $x \in X$.

We will use the *vector norm*

(2.6) $$\|X\| = \max (|X_1|, \cdots, |X_n|)$$

for interval vectors $X = (X_1, \cdots, X_n)$.

It is useful to define the *midpoint* of an interval

(2.7) $$m(X) = (\underline{X} + \bar{X})/2.$$

Similarly, if X is an interval vector, we write $m(X) = (m(X_1), \cdots, m(X_n))$.

An *interval matrix* is a matrix whose coefficients are intervals. If A is an interval matrix with coefficients A_{ij} and if B is a matrix with real number coefficients B_{ij} such that $B_{ij} \in A_{ij}$ for all i and j, then we write $B \in A$. We will use the *matrix norm*

(2.8) $$\|A\| = \max_i \sum_j |A_{ij}|$$

for an interval matrix A. This is an interval extension of the maximum row sum norm for real matrices. Note that if B is any real matrix contained in an interval matrix A, then $\|B\| \leq \|A\|$. We can compute $\|A\|$ using (2.8) and (2.5).

We will also use the notation

(2.9) $$w(A) = \max_{i,j} w(A_{ij}) \quad \text{(the width of the interval matrix } A\text{)}$$

and

(2.10) $$(m(A))_{ij} = m(A_{ij}) \quad (m(A) \text{ is the } \textit{midpoint matrix } \text{in } A).$$

The definition (2.10) means that we define a matrix $m(A)$ with real coefficients $(m(A))_{ij}$ chosen as the midpoints of the corresponding coefficients of the interval matrix A. Clearly, $m(A) \in A$.

2.2. Interval arithmetic. We can treat intervals X and Y as numbers, adding them as follows. $X + Y = Z$, where $\underline{Z} = \underline{X} + \underline{Y}$ and $\bar{Z} = \bar{X} + \bar{Y}$. Put another way, we can add the inequalities

$$\underline{X} \leq x \leq \bar{X} \quad \text{and} \quad \underline{Y} \leq y \leq \bar{Y}$$

to obtain $\underline{X} + \underline{Y} \leq x + y \leq \bar{X} + \bar{Y}$. Thus, we can compute the *set*

(2.11) $$X + Y = \{x + y : x \in X, y \in Y\}.$$

Thus, the sum of two intervals is again an interval. It is the interval of sums of real numbers, one from the first interval and the other from the second interval. Similarly, we define the negative of an interval by

(2.12) $$-X = -[\underline{X}, \bar{X}] = [-\underline{X}, -\underline{X}] = \{-x : x \in X\}.$$

For the difference of two intervals, we form

(2.13) $$Y - X = Y + (-X) = \{y - x : x \in X, y \in Y\}.$$

More briefly, the rules for *interval addition* and *subtraction* are:

(2.14) $$[\underline{X}, \bar{X}] + [\underline{Y}, \bar{Y}] = [\underline{X} + \underline{Y}, \bar{X} + \bar{Y}]$$

(2.15) $$[\underline{X}, \bar{X}] - [\underline{Y}, \bar{Y}] = [\underline{X} - \bar{Y}, \bar{X} - \underline{Y}].$$

We can define the *reciprocal* of an interval as follows

(2.16) $$1/X = \{1/x : x \in X\}.$$

If X is an interval *not* containing the number 0, then

(2.17) $1/X = [1/\bar{X}, 1/\underline{X}]$.

In other words, (2.17) yields (2.16) for $X = [\underline{X}, \bar{X}]$ provided that either $\underline{X} > 0$ or $\bar{X} < 0$.

If X contains 0, so that $\underline{X} \leqq 0 \leqq \bar{X}$, then $x \in X$ implies that $1/x \geqq 1/\bar{X}$ or (but not *and*) $1/x \leqq 1/\underline{X}$. In this case the set (2.16) is unbounded and cannot be represented as an interval whose endpoints are real numbers. (N 1).

For the *product* of two intervals, we define

(2.18) $X \cdot Y = \{xy : x \in X, y \in Y\}$.

It is not hard to see that $X \cdot Y$ is again an interval, whose endpoints can be computed from

(2.19)
$$\underline{X \cdot Y} = \min(\underline{X}\underline{Y}, \underline{X}\bar{Y}, \bar{X}\underline{Y}, \bar{X}\bar{Y})$$
$$\overline{X \cdot Y} = \max(\underline{X}\underline{Y}, \underline{X}\bar{Y}, \bar{X}\underline{Y}, \bar{X}\bar{Y}).$$

By testing the signs of the endpoints of X and Y, we can reduce the formulas (2.19) to nine special cases. In eight of these, only one real product is needed for each end point of $X \cdot Y$. The ninth case requires two real products for each endpoint of $X \cdot Y$.

(2.20)

(1) $0 \leqq \underline{X}, 0 \leqq \underline{Y}$: $\underline{X \cdot Y} = \underline{X}\underline{Y}, \overline{X \cdot Y} = \bar{X}\bar{Y}$,

(2) $\underline{X} < 0 < \bar{X}, 0 \leqq \underline{Y}$: $\underline{X \cdot Y} = \underline{X}\bar{Y}, \overline{X \cdot Y} = \bar{X}\bar{Y}$,

(3) $\bar{X} \leqq 0, 0 \leqq \underline{Y}$: $\underline{X \cdot Y} = \underline{X}\bar{Y}, \overline{X \cdot Y} = \bar{X}\underline{Y}$,

(4) $0 \leqq \underline{X}, \underline{Y} < 0 < \bar{Y}$: $\underline{X \cdot Y} = \bar{X}\underline{Y}, \overline{X \cdot Y} = \bar{X}\bar{Y}$,

(5) $\bar{X} \leqq 0, \underline{Y} < 0 < \bar{Y}$: $\underline{X \cdot Y} = \underline{X}\bar{Y}, \overline{X \cdot Y} = \underline{X}\underline{Y}$,

(6) $0 \leqq \underline{X}, \bar{Y} \leqq 0$: $\underline{X \cdot Y} = \bar{X}\underline{Y}, \overline{X \cdot Y} = \underline{X}\bar{Y}$,

(7) $\underline{X} < 0 < \bar{X}, \bar{Y} \leqq 0$: $\underline{X \cdot Y} = \bar{X}\underline{Y}, \overline{X \cdot Y} = \underline{X}\underline{Y}$,

(8) $\bar{X} \leqq 0, \bar{Y} \leqq 0$: $\underline{X \cdot Y} = \bar{X}\bar{Y}, \overline{X \cdot Y} = \underline{X}\underline{Y}$,

(9) $\underline{X} < 0 < \bar{X}, \underline{Y} < 0 < \bar{Y}$: for this case, we have

$$\underline{X \cdot Y} = \min(\underline{X}\bar{Y}, \bar{X}\underline{Y})$$
$$\overline{X \cdot Y} = \max(\underline{X}\underline{Y}, \bar{X}\bar{Y}).$$

For the quotient of two intervals, we define

(2.21) $X/Y = X \cdot (1/Y) = \{x/y : x \in X, y \in Y\}$.

Recall that $1/Y$ is defined by (2.16). If 0 is not contained in Y, then X/Y is again an interval and its endpoints can be computed using (2.17) and (2.19) or (2.20). Thus, X/Y is an interval if $\underline{Y} > 0$ or $\bar{Y} < 0$.

For brevity, we will sometimes drop the dot notation for the product of two intervals and simply write XY for the product of X and Y.

2.3. Algebraic properties of interval arithmetic. The following algebraic properties of interval arithmetic are immediate consequences of the set theoretic definitions of the interval arithmetic operations (2.11), (2.13), (2.18), (2.21):

(2.22)
$$X + (Y + Z) = (X + Y) + Z,$$
$$X(YZ) = (XY)Z,$$
$$X + Y = Y + X,$$
$$XY = YX$$

for any intervals X, Y, Z.

In addition and multiplication we have

(2.23)
$$0 + X = X + 0 = X, \qquad 0X = X0 = 0,$$
$$1X = X1 = X$$

for any interval X.

Thus, addition and multiplication are *associative* and *commutative*. However, the *distributive* law does *not* always hold. For example, we have

$$[1, 2] \cdot (1 - 1) = 0;$$

whereas

$$[1, 2] \cdot 1 - [1, 2] \cdot 1 = [-1, 1] \neq 0.$$

Thus, $X(Y + Z) = XY + XZ$ is *not* always true.

We do, however, always have the following algebraic property

(2.24) $$X(Y + Z) \subseteq XY + XZ.$$

We call this property *subdistributivity*. As can be seen, it is really a combination of algebraic *and* set theoretic relations. (N 5).

In certain special cases, distributivity holds. Some particularly useful cases are:

(2.25)
$$x(Y + Z) = xY + xZ \qquad \text{for } x \text{ real}; Y, Z \text{ intervals},$$
$$X(Y + Z) = XY + XZ \quad \text{if } YZ > 0.$$

Thus, we can distribute multiplication by a real number and we can distribute multiplication by any interval through sums of intervals all of the same sign.

The properties (2.24) and (2.25) follow easily from the definitions of interval arithmetic.

With the identification of degenerate intervals and real numbers, interval arithmetic is an extension of real arithmetic and reduces to ordinary real arithmetic for intervals of zero width.

Note that $X - X = 0$ and $X/X = 1$ *only* when X is of width zero. Otherwise, $X - X = [\underline{X} - \bar{X}, \bar{X} - \underline{X}] = w(X)[-1, 1]$; and $X/X = [\underline{X}/\bar{X}, \bar{X}/\underline{X}]$ for $0 < \underline{X}$; and $X/X = [\bar{X}/\underline{X}, \underline{X}/\bar{X}]$ for $\bar{X} < 0$. The *cancellation law* holds for interval addition:

$$X + Z = Y + Z$$

implies $X = Y$.

By a *symmetric* interval, we mean an interval X such that $\underline{X} = -\bar{X}$. Thus, X is symmetric if and only if $m(X) = 0$. For a symmetric interval X, we have $|X| = w(X)/2$ and $X = |X|[-1, 1]$. If X and Y are symmetric intervals, then $X + Y = X - Y = (|X| + |Y|)[-1, 1]$; furthermore, $XY = |X||Y|[-1, 1]$. If X, Y, and Z are symmetric intervals, then $X(Y \pm Z) = XY + XZ = |X|(|Y| + |Z|)[-1, 1]$.

An arbitrary interval X can be written as the sum of a real number and a symmetric interval. Thus,

(2.26) $X = m + W$,   where $m = m(X)$ and $W = \frac{1}{2}w(X)[-1, 1]$.

Put another way, we can write

(2.27) $X = [\underline{X}, \bar{X}] = (\underline{X} + \bar{X})/2 + ((\bar{X} - \underline{X})/2)[-1, 1]$.

If Y is a symmetric interval and X is any interval, then $XY = |X|Y$. It follows that $X(Y + Z) = XY + XZ$ if Y and Z are symmetric, for *any* interval X. (N 2), (N 3), (N 4), (N 6).

2.4.   Rounded interval arithmetic. Nearly all numerical computation is carried out with "fixed-precision", approximate arithmetic. We decide, usually in advance of beginning a computation, how many digits (or bits) we are going to carry and we truncate all intermediate and final results to that many digits. The arithmetic hardware of computers is designed to carry out approximate arithmetic in "fixed-precision". Numbers are represented in the computer by strings of bits of fixed, *finite* length. Most commonly, so-called floating point arithmetic is used, in which numbers are represented in the computer by a string of bits of fixed length, $b_0 b_1 \cdots b_s$, s fixed.

Such a string of bits (each b_i is 0 or 1) is taken to represent a "floating point" number of the form $m \cdot 2^e$, where the "mantissa" m, is represented by s_1 bits and e is represented by $s + 1 - s_1$ bits. Here, e will be a signed integer, in some allowed range, and m may be zero or a signed binary fraction with $s_1 - 1$ bits plus sign. It is possible to program a computer to carry out the operations of interval arithmetic with appropriate *rounding*, when necessary, of left and right computed endpoints, so that the machine computed interval result *always contains* the exact interval result, or else the computer indicates that it cannot do so (in case of overflow, for instance, when the allowed range of exponents, e, would be exceeded).

In some cases, this requires adding (or subtracting) a "low order bit" to (or from) the right (or left) hand endpoint of a machine computed interval result. This can be done in such a way that the machine computed interval result not only contains the exact interval result, but the machine computed right endpoint is the smallest[1] machine number not less than the correct right endpoint and the machine computed left endpoint is the largest machine number not greater than the correct left endpoint. This is called "best possible" *rounded interval arithmetic*. (N 9).

Computer realization of interval arithmetic can be accomplished in various ways (N 10). Algol or FORTRAN compilers can be augmented so that a variable of "interval type" is recognized and the appropriate arithmetic operations are compiled for expressions involving that variable (N 8). This can be accomplished either with subroutine or procedure calls to subprograms written in Algol or FORTRAN or it can be accomplished by *microprogramming* the computer in advance, so that interval arithmetic operations (properly rounded) are defined at the machine language level.

The advantage of defining interval arithmetic on a computer via user microprogrammable (writeable) control stores is one of efficiency. Subroutine calls will be necessarily slow compared to hardware arithmetic. By use of microprogramming, rounded interval arithmetic can be performed at about the same speed as ordinary hardware arithmetic. There is an initial cost of achieving such higher efficiency, namely the time required to produce the required microprogrammed rounded interval arithmetic in the control store of a particular machine. It is likely that this will be made easier in the coming years.

Without any "system" at all (subroutines, microprogramming or otherwise) rounded interval arithmetic can be programmed for any specific application by programming separately the computation of left and right endpoints of intervals. To take the error in finite-precision computer arithmetic into account, all that is necessary is to multiply positive right endpoints of computed results by a factor of the form $(1 + 2^{-t})$ for an appropriate integer t, depending on the number s of "significant" bits $s > t$ carried in the mantissa of numbers for that particular machine representation, and to multiply negative left endpoints, similarly, by $(1 + 2^{-t})$ (N 7). For large enough s and t, we can make rounded interval arithmetic as close as we please to exact interval arithmetic.

A numerical example will illustrate the distinction between interval arithmetic and rounded interval arithmetic.

---

[1] With respect to the order relation $\leq$.

Let
$$X = [-.613 \cdot 10^{-2}, -.610 \cdot 10^{-2}],$$
$$Y = [+.100 \cdot 10^{+1}, +.300 \cdot 10^{+1}],$$
$$Z = X(1 + 1/Y).$$

We compute Z first in exact interval arithmetic. We have

$$Z = X(1 + 1/[1, 3])$$
$$= X(1 + [1/3, 1])$$
$$= X[4/3, 2]$$
$$= [-.1226 \cdot 10^{-1}, -.81333 \ldots \cdot 10^{-2}].$$

Next, we compute Z using rounded interval arithmetic based on signed three decimal digit mantissas and floating point number representation. We have

$$1 = [+.100 \cdot 10^{+1}, +.100 \cdot 10^{+1}],$$
$$1/Y \subseteq [+.333 \cdot 10^{+0}, +.100 \cdot 10^{+1}],$$
$$1 + 1/Y \subseteq [+.133 \cdot 10^{+1}, +.200 \cdot 10^{+1}],$$
$$X[+.133 \cdot 10^{+1}, +.200 \cdot 10^{+1}] \subseteq [-.123 \cdot 10^{-1}, -.811 \cdot 10^{-2}].$$

Each of the four numerical intervals on the right hand sides of the above relations is the smallest (narrowest) interval representable in the chosen number form containing the quantity on the left. The final result contains the exact interval value of Z. It is not the narrowest interval of the chosen form which contains the exact value of Z, because of the cumulative effect of dropping digits beyond the third one after the decimal point. By carrying enough digits we can come as close as we please to the exact interval arithmetic results.

2.5. Functions, algorithms, computer programs. Finite representations of functions are possible in a variety of ways. We can store and "look up" a finite set of pairs of argument and function values. We can use piecewise-linear or other interpolation procedures for intermediate values. We can program formulas and algorithms for computing values in finite sets of computer words, representing machine executable instructions for carrying out arithmetic and search operations resulting in function values. In short, *computer programs* provide a means for finite representations of functions. In particular, we can write computer programs which use rounded interval arithmetic and which operate with interval valued arguments and interval valued functions. In the remaining chapters, we will study how this can be done for a variety of useful purposes.

NOTES

1. The definition and implementation of an arithmetic for intervals on the extended real line, including *unbounded intervals*, has been reported by S. E. Laveuve [70, pp. 236–245] based on ideas of W. M. Kahan.

2. F. N. Ris introduces a number of real valued functions on intervals (in addition to $m(X)$, $w(X), \cdots$) and discusses their use in analyzing interval algorithms [70, pp. 75–98].

3. An arithmetic for *circular discs* in the *complex plane* has been developed and applied to the computation of complex zeros of polynomials and to the theory of continued fractions; see P. Henrici [70, pp. 14–30], I. Gargantini and P. Henrici [17], G. Glatz [70, p. 205–214], I. Gargantini [70, pp. 196–204], [16], [15a], N. Krier [44], K. Nickel [25, pp. 25–34], G. Alefeld and J. Herzberger [1a], N. Krier and P. Spellucci [70, pp. 223–228].

4. A number of investigations have been undertaken concerning the possibilities for doing arithmetic with *probability distributions*; see R. Ahmad [70, pp. 127–134], M. Dempster [70, pp. 107–127], K. Ecker and H. Ratschek [15].

5. For further discussion of *distributivity* in interval arithmetic; see H. Ratschek [79], O. Spaniol [90].

6. Other types of arithmetic with sets in n dimensions have been considered (in particular for *ellipsoids* and *polytopes*); see K. G. Guderley and C. L. Keller [22], L. W. Jackson [36], W. M. Kahan [40], D. P. Davey and N. F. Stewart [12], N. F. Stewart [91].

7. For an example of the *direct programming in FORTRAN* of rounded interval arithmetic operations, see R. E. Moore [59, p. 15].

8. A number of *programs* or *programming systems* for computing with rounded interval arithmetic have been published. The most extensively developed of these are based on the algorithmic language "*triplex-Algol 60*"; see N. Apostolatos, U. Kulisch, R. Krawczyk, B. Lortz, K. Nickel, and H.-W. Wippermann [4]. A *triplex* number is an ordered triple of machine numbers $[\underline{x}, \tilde{x}, \bar{x}]$ satisfying $\underline{x} \leq \tilde{x} \leq \bar{x}$. Thus, $[\underline{x}, \bar{x}]$ is an interval and $\tilde{x}$ is some real number in that interval; $\tilde{x}$ is called the "main" value of the triplex number. For a discussion of some of the advantages of the triplex representation of intervals, see K. Nickel [25, pp. 10–24]. For a realization of rounded interval arithmetic on an *arbitrary Algol 60 compiler*; see H. Christ [9]. For computer realizations based on *FORTRAN subroutines*, see J. M. Yohe [100]. The system for performing interval computation in FORTRAN described by Yohe has been implemented on UNIVAC, CDC, DEC, Honeywell, and IBM equipment. Triplex-Algol 60 has also been implemented on several computers, including: Zuse Z 23, Electrologica X8, and UNIVAC 1108, see preface in K. Nickel [70]. Many of the papers listed in the references contain *Algol programs* for interval algorithms.

9. In a recent fundamental work, U. Kulisch has made a thorough study of the mathematical foundations of numerical computing, machine arithmetic, rounding procedures, and rounded interval arithmetic [48]. See also H. Apostolatos and U. Kulisch [5].

10. See J. M. Yohe [100] and also D. I. Good and R. L. London [18]; for an approach to error arithmetic based on relative error, see F. W. J. Olver [73].

Chapter **3**

Finite Evaluation

3.1. Set valued mappings. Let M_1 and M_2 be *arbitrary sets* and let $g: M_1 \to M_2$ be an *arbitrary mapping* (function) from M_1 into M_2. Denote by $S(M_1)$ and $S(M_2)$ the families of subsets of M_1 and M_2 respectively. Following W. Strother (N 1), we call the *set-valued mapping*, $\bar{g}: S(M_1) \to S(M_2)$,

(3.1) $$\bar{g}(X) = \{g(x): x \in X, \, X \in S(M_1)\}$$

the *united extension* of g. We can also write

$$\bar{g}(X) = \bigcup_{x \in X} \{g(x)\}.$$

Thus, $\bar{g}(X)$ is the union in M_2 of all the sets containing a single element $g(x)$ for some x in X. Sometimes, $\bar{g}(X)$ is simply referred to as "the image under the mapping g of the set X". We prefer the notation shown above to the commonly used $g(X)$ because we want to study $\bar{g}$ as a *single valued* mapping on $S(M_1)$ with values in $S(M_2)$.

Now, even if g has a finite representation, say as a polynomial, and X has a finite representation, say as an interval, we will still, *in general, not* have a finite representation for $\bar{g}$. We will not be able to compute, in a finite number of arithmetic operations, the interval value $\bar{g}(X)$. (See Example following Corollary 3.1: equation (3.7) and following paragraphs.)

On the other hand, there are *some* set valued mappings which *do* have finite representations and which *can* be evaluated in a finite number of arithmetic operations. This is the case, in particular, for *rational interval functions*.

The endpoint formulas (2.14), (2.15), (2.17), and (2.20) for the interval arithmetic operations enable us to compute the united extensions of real arithmetic functions for pairs of intervals as defined by (2.11), (2.13), (2.18), and (2.21).

By a "rational interval function", we mean a function whose *interval values* are defined by a specific finite sequence of interval arithmetic operations. For

19

example, consider the mapping F whose values are defined by

$$T_1 = [1, 2] X_1,$$

(3.2) $$T_2 = T_1 + [0, 1],$$

$$F(X_1, X_2) = T_2 X_2.$$

for intervals X_1, and X_2.

Here, F is a finitely represented mapping from the set of all pairs of intervals (X_1, X_2) into the set of intervals. It can be looked at as a set valued mapping, whose *values* are *sets* of real numbers, namely: closed bounded intervals of real numbers. For this particular rational interval function, we will have $F(X_1, X_2) = \bar{g}([1, 2], [0, 1], X_1, X_2)$ where $g(c_1, c_2, x_1, x_2) = (c_1 x_1 + c_2) x_2$ is real valued with real arguments. The formula $G(X_1, X_2) = X_1 X_2$ defines another rational interval function. Taken together, as the first and second components of an interval vector mapping, F and G map two-dimensional rectangles into two-dimensional rectangles. For instance, the interval vector $(X_1, X_2) = ([0, 1], [0, 1])$ is mapped into the interval vector $([0, 3], [0, 1])$. Not all rational interval functions are united extensions of real functions. For example, $F(X) = X - X$ is not a united extension of any $f(x)$, since $F(x) = 0$ for all degenerate intervals x, but we do not have $F(X) = 0$ for all X.

3.2. The subset property and inclusion monotonicity. Again, let $g: M_1 \to M_2$ be an *arbitrary* mapping, M_1 and M_2 *arbitrary* sets. The united extension, $\bar{g}: S(M_1) \to S(M_2)$, has the *subset property*:

(3.3) $$X, Y \in S(M_1) \text{ with } X \subseteq Y \quad \text{implies} \quad \bar{g}(X) \subseteq \bar{g}(Y).$$

The subset property (3.3) follows directly from the definition (3.1) of the united extension.

It is important for many applications of interval computation that interval extensions (§ 3.3), in addition to those which are united extensions of real functions, have a similar property. We say that an interval valued function F of the interval variables $X_1, X_2, \cdots, X_n$ is *inclusion monotonic* if

$$Y_i \subseteq X_i, \qquad i = 1, 2, \cdots, n,$$

implies

(3.4) $$F(Y_1, Y_2, \cdots, Y_n) \subseteq F(X_1, X_2, \cdots, X_n).$$

United extensions, which *all have the subset property, are inclusion monotonic.* Since the interval arithmetic functions are united extensions of the real arithmetic functions $(+, -, \cdot, /)$, we have the result that *interval arithmetic is inclusion monotonic*:

$$Y_1 \subseteq X_1 \quad \text{and} \quad Y_2 \subseteq X_2$$

implies

$$Y_1 + Y_2 \subseteq X_1 + X_2,$$

(3.5) $$Y_1 - Y_2 \subseteq X_1 - X_2,$$

$$Y_1 Y_2 \subseteq X_1 X_2,$$

$$Y_1 / Y_2 \subseteq X_1 / X_2.$$

From the transitivity of the partial order relation, $\subseteq$, it follows, by finite induction, that *rational interval functions are inclusion monotonic, as are natural interval extensions of all the standard functions used in computing,* (see § 3.3, § 4.4). With proper rounding procedures, *rounded interval arithmetic operations are also inclusion monotonic* (N 2).

Not all interval valued functions, however, are inclusion monotonic. For example, consider the interval function F defined by

$$F(X) = m(X) + \tfrac{1}{2}(X - m(X)).$$

We have $F([0, 2]) = 1 + \tfrac{1}{2}[-1, 1] = [\tfrac{1}{2}, \tfrac{3}{2}]$; whereas $F([0, 1]) = \tfrac{1}{2} + \tfrac{1}{2}[-\tfrac{1}{2}, \tfrac{1}{2}] = [\tfrac{1}{4}, \tfrac{3}{4}] \not\subseteq F([0, 2])$. This interval valued function does *not* arise as an interval extension of a real valued function. Its restriction to real x is the identity mapping.

3.3. Interval extensions. Let f be a real valued function of n real variables $x_1, \cdots, x_n$. By an *interval extension* of f, we mean an interval valued function F of n interval variables $X_1, \cdots, X_n$ with the property

(3.6) $$F(x_1, \cdots, x_n) = f(x_1, \cdots, x_n), \quad \text{for real arguments.}$$

Thus, an interval extension of f is an interval valued function which has real values when the arguments are all real (degenerate intervals) and coincides with f in this case.

THEOREM 3.1. *If F is an inclusion monotonic interval extension of f, then $\bar{f}(X_1, \cdots, X_n) \subseteq F(X_1, \cdots, X_n)$.*

Proof. By the definition of an interval extension, $f(x_1, \cdots, x_n) = F(x_1, \cdots, x_n)$. If F is inclusion monotonic, then the value of f is contained in the interval $F(X_1, \cdots, X_n)$ for every $(x_1, \cdots, x_n)$ in $(X_1, \cdots, X_n)$.

There is never a unique interval extension of a given real function. For instance, if F is an interval extension of f such that $F(x) = f(x)$ for real x, then $F_1(X) = F(X) + X - X$ defines another, different, interval extension of f.

Real rational functions of n real variables have *natural* interval extensions. Given a rational expression in real variables, we can replace the real variables by corresponding interval variables and replace the real arithmetic operations by the corresponding interval arithmetic operations to obtain a rational interval function which is a natural extension of the real rational function. On the other hand, two rational expressions which are equivalent in real arithmetic may not be equivalent in interval arithmetic. They may give rise to two different rational interval functions. For example, let $f(x) = x(1-x) = x - x \cdot x$. For the first expression, we obtain the interval extension $F_1(X) = X(1-X)$. For the second, we obtain $F_2(X) = X - X \cdot X$. Now we have $F_1([0, 1]) = [0, 1] \, (1 - [0, 1]) = [0, 1]$; whereas $F_2([0, 1]) = [0, 1] - [0, 1] \cdot [0, 1] = [-1, 1]$.

Since rational interval functions are inclusion monotonic, we have the following corollary to Theorem 3.1.

COROLLARY 3.1. *If F is a rational interval function and an interval extension of f, then $\bar{f}(X_1, \cdots, X_n) \subseteq F(X_1, \cdots, X_n)$. In other words, an interval value of F* contains the range of values *of the corresponding real function f (f is the "real restriction" of F) when the real arguments of f lie in the intervals shown.*

Corollary 3.1 provides us with a means for the finite evaluation of upper and lower bounds on the ranges of values of real rational functions over n dimensional rectangles. For example, consider the polynomial

$$(3.7) \qquad\qquad p(x) = 1 - 5x + \tfrac{1}{3}x^3.$$

Suppose we wish to know the range of values of $p(x)$ when x is any number in the interval $[2, 3]$. A natural interval extension of p is the interval polynomial

$$(3.8) \qquad\qquad P(X) = 1 - 5X + \tfrac{1}{3}X \cdot X \cdot X.$$

Computing $P([2, 3])$, we obtain

$$P([2, 3]) = 1 - 5[2, 3] + \tfrac{1}{3}[8, 27] = [-(\tfrac{34}{3}), 0].$$

Thus, we have found, by a finite sequence of arithmetic computations, that the range of values of $p(x)$ when x is in $[2, 3]$ is contained in the interval $[-\tfrac{34}{3}, 0]$. Another natural extension we can use is as follows. Rewrite $p(x)$ in the form

$$p(x) = 1 - x(5 - x^2/3),$$

which is equivalent in real arithmetic to the original form. A natural interval extension of $p(x)$ written in this form is

$$(3.9) \qquad\qquad Q(X) = 1 - X(5 - X \cdot X/3).$$

Computing $Q([2, 3])$, we obtain

$$\begin{aligned}
Q([2, 3]) &= 1 - [2, 3](5 - [2, 3] \cdot [2, 3]/3) \\
&= 1 - [2, 3](5 - [\tfrac{4}{3}, 3]) \\
&= 1 - [2, 3][2, \tfrac{11}{3}] \\
&= 1 - [4, 11] \\
&= [-10, -3].
\end{aligned}$$

Thus, we have found a narrower interval which also contains the range of values of $p(x)$ for x in $[2, 3]$.

The exact range of values for this $p(x)$ for x in $[2, 3]$ is $[-\frac{10}{3}\sqrt{5}+1, -5]$ and *cannot* be found exactly by *any* finite sequence of arithmetic operations with rational numbers, since $\sqrt{5}$ is an irrational number. In the next chapter we will discuss some interval methods for computing convergent sequences of upper and lower bounds to exact ranges of values.

In the example above, we found upper and lower bounds on the range of values of a polynomial by *evaluating* an interval polynomial using interval arithmetic. The bounds we obtain depend on which interval extension we use. Using $Q(X)$ in (3.9) we obtained

$$-10 \leqq p(x) \leqq -3 \quad \text{for all } x \text{ in } [2, 3]$$

for $p(x)$ given by (3.7).

For polynomials, the nested form (sometimes called Horner's scheme or synthetic division), $A_0 + X(A_1 + X(A_2 + \cdots + X(A_n) \cdots)$ is never worse than and usually better than the sum of powers $A_0 + A_1 X + A_2 X \cdot X + \cdots + A_n X \cdot X \cdot \cdots \cdot X$, because of subdistributivity.

From the set theoretic definitions of interval arithmetic it follows that *any natural interval extension of a rational function in which each variable occurs only once* (if at all) and to the first power only *will compute the exact range of values* providing that no division by an interval containing zero occurs.

We can *construct inclusion monotonic interval extensions* of other commonly used functions, including all standard FORTRAN functions (§4.4).

For positive integer values of n, we define the *powers of intervals* by

$$(3.10) \qquad X^n = \begin{cases} [\underline{X}^n, \bar{X}^n] & \text{if } \underline{X} > 0 \text{ or } n \text{ is odd;} \\ [\bar{X}^n, \underline{X}^n] & \text{if } \bar{X} < 0 \text{ and } n \text{ is even,} \\ [0, |X|^n] & \text{if } 0 \in X \text{ and } n \text{ is even.} \end{cases}$$

We have $X^n = \{x^n : x \in X\}$. To take the limited precision of machine arithmetic into account, we can treat $\underline{X}$, $\bar{X}$, and $|X|$ as degenerate intervals and compute $\underline{X}^n, \bar{X}^n$, and $|X|$ in rounded interval arithmetic, using the appropriate endpoint obtained in this way in the above formulas for X^n in the various cases. Thus we can compute, in each case, a machine interval containing the exact result.

For *monotonic increasing functions* $f(x)$ such as $\sqrt{x}$, $\exp x$, $\log_e x$ we have $\bar{f}(X) = [f(\underline{X}), f(\bar{X})]$. We could use the united extensions of such functions except that we cannot usually compute the required endpoint value exactly. However, we can extend the machine-computed endpoints outward in each case by an amount just great enough to include the approximation error for each such function. In this way we can compute, on the machine, intervals containing the exact ranges of values. Monotonic decreasing functions can be treated in a similar way. Interval extensions *constructed* in this way, are *inclusion montonic.*

For functions which are not monotonic (in the ordinary sense) such as sin x, we can use other techniques to obtain appropriate interval extensions. This can be done, for instance, for *all standard FORTRAN functions* (N 3). It can be done in such a way that special properties of the functions are preserved by the inclusion monotonic interval extensions as well. For example, we can ensure that the interval extension we use for the sine function does not give values outside the interval $[-1, +1]$ by *intersecting* this interval with a computed interval $F(X)$ bounding the range of values of sin x for x in some interval X. The resulting interval, $F(X) \cap [-1, 1]$, will still be an *inclusion* monotonic interval extension of sin x and will not have values outside $[-1, 1]$.

For real valued functions defined in a "piece-wise" manner, we can *construct* inclusion monotonic interval extensions as in the following illustration. Suppose that f is defined by

$$(3.11) \qquad f(x) = \begin{cases} p_1(x) & \text{for } x_1 \leqq x < x_2, \\ p_2(x) & \text{for } x_2 \leqq x < x_3, \\ p_3(x) & \text{for } x_3 \leqq x \leqq x_4. \end{cases}$$

And suppose we have interval extensions P_1, P_2, P_3 for p_1, p_2, p_3. We can *construct* an inclusion monotonic interval extension F for f as follows:

$$(3.12) \quad F(X) = P_1(X \cap [x_1, x_2]) \cup P_2(X \cap [x_2, x_3]) \cup P_3(X \cap [x_3, x_4]).$$

In this way we can compute bounds on the range of values, for instance, of spline functions when the argument x lies in an interval X.

3.4. Recursive evaluation of derivatives. Computer programs for the *efficient* evaluation of Taylor series coefficients have been available for more than a decade (N 4). The techniques used have evidently still not reached a wide audience. One still sees, in publication after publication, statements to the effect that "the direct use of Taylor series expansions, for the numerical solution of the initial value problem in ordinary differential equations is not of practical value because of the difficulty of obtaining Taylor coefficients". The exact opposite is true, if we use the right approach (N 5). (See also § 8.1)

The technique to be described here is not that of formula manipulation by computer. We do *not* need to *see formulas* for successive Taylor coefficients in order to be able to evaluate successive derivatives of a function at some point. In fact, *formulas*, fully written out, for successive derivatives usually *do* grow in complexity extremely fast; but they are *not needed.*

Instead, we can operate on a computer with an approach which amounts to *differentiating subroutines* (or function subprograms). The result of differentiating a program is another program. The *execution* of the derived program produces a value of the derivative of the function defined by the first program. In fact, we can write a general purpose program which, when applied to a function defined by a (say FORTRAN) program in a computer, will produce a Kth derivative program for general K for that function. If we supply it with an integer K and argument values, it will produce values of the derivatives (or Taylor coefficients) up to order K of the given function at the given argument values. This can be carried out either in machine arithmetic or in rounded interval arithmetic (N 4). Partial derivatives can also be obtained in a similar way (N 4).

It is useful to introduce the following notation. Suppose $x(t)$ is analytic in t in some neighborhood of t_0. We define

(3.13)
$$(x)_0 = x(t_0),$$
$$(x)_k = (1/k!)(d^k x/dt^k)(t_0), \qquad k = 1, 2, \cdots.$$

Thus the notation $(x)_k$ stands for the kth Taylor coefficient in the expansion of $x(t)$ about $t = t_0$:

(3.14)
$$x(t) = \sum_{k=0}^{\infty} (x)_k (t - t_0)^k.$$

For the finite Taylor series expansion with mean value form of remainder, we have

(3.15)
$$x(t) = \sum_{k=0}^{N-1} (x)_k (t - t_0)^k + r_N (t - t_0)^N,$$

If R_N is an interval extension of $r_N(s)$, so that $r_N(s) \in R_N([t_0, t])$, for every s in $[t_0, t]$, then for t within the radius of convergence of the Taylor series (3.14) we will have

(3.16)
$$x(t) \in \sum_{k=0}^{N-1} (x)_k (t - t_0)^k + R_N([t_0, t])(t - t_0)^N,$$

where the expression on the right hand side is an interval valued function containing the Taylor series with remainder (3.15).

We will now show how to compute the real and interval values $(x)_k$ and $R_N([t_0, t])$ for almost any function x which will ever appear in a computer program.

From the definition (3.13) we find that

(3.17)
$$x'(t_0) = (x)_1 \quad \text{and so} \quad (x)_k = (1/k)((x)_1)_{k-1}.$$

This will be an important relation, because if we have a function $x(t)$ defined by a differential equation, for instance,

$$x'(t) = g(t, x(t)) \quad \text{with a given } x(t_0),$$

then we can use $(x)_k = (1/k)(g)_{k-1}$ recursively to compute the Taylor coefficients for $x(t)$ about t_0, provided—of course—that we can handle $g(t, x(t))$. Similarly, $(x)_k = (2/(k(k-1)))((x)_2)_{k-2}$.

Let u and v be analytic functions of t in some neighborhood of $t = t_0$. For arithmetic functions of u and v we have:

(3.18)

$$(u+v)_k = (u)_k + (v)_k,$$

$$(u-v)_k = (u)_k - (v)_k,$$

$$(uv)_k = \sum_{j=0}^{k} (u)_j(v)_{k-j},$$

$$(u/v)_k = (1/v)\left\{(u)_k - \sum_{j=1}^{k} (v)_j(u/v)_{k-j}\right\}.$$

The derivation of the formulas (3.18) is a simple exercise in elementary calculus.

All the commonly used "elementary" functions satisfy *rational* differential equations. Some of these are second order differential equations, which we can rewrite as a pair of first order differential equations. This, together with the relations (3.17) enables us to derive *recursion relations* for the kth Taylor coefficients of elementary functions of an arbitrary analytic function $u(t)$. For example, we can derive, in this way, the following recursion relations:

$$(u^a)_k = (1/u) \sum_{j=0}^{k-1} (a - j(a+1)/k)(u)_{k-j}(u^a)_j,$$

$$(e^u)_k = \sum_{j=0}^{k-1} (1 - j/k)(e^u)_j(u)_{k-j},$$

(3.19)

$$(\log_e u)_k = (1/u)\left((u)_k - \sum_{j=1}^{k-1} (1 - j/k)(u)_j(\log_e u)_{k-j}\right),$$

$$(\sin u)_k = (1/k) \sum_{j=0}^{k-1} (j+1)(\cos u)_{k-1-j}(u)_{j+1},$$

$$(\cos u)_k = -(1/k) \sum_{j=0}^{k-1} (j+1)(\sin u)_{k-1-j}(u)_{j+1}.$$

The "chain rule", $df(u)/dt = f'(u)u'(t)$ is also useful in the course of deriving such recursion relations. Similar recursion relations can be derived for hyberbolic functions, Bessel functions, and all the other commonly used functions in applied mathematics.

Sometimes the rules simplify; for example, if u is a constant (independent of t), then $(uv)_k = u(v)_k$ because $(u)_j = 0$ for $j \geq 1$. As another example, if $u(t) = t$, then $(e^u)_k$ becomes $(e^t)_k = (1/k)(e^t)_{k-1}$ because $(t)_{k-j} = 0$ for $j < k-1$, and $(t)_1 = 1$.

A general procedure for applying these rules to the computation of Taylor coefficients of functions $x(t)$ can be programmed to operate as follows.

1. Represent $x(t)$ (or $x'(t)$ or $x''(t) \cdots$) by a finite list of binary or unary operations (e.g. $T_3 = T_1 + T_2$, $T_4 = e^{T_3}$, etc.);
2. On a line-by-line basis, generate subprograms for Taylor coefficients for each item in the list, using the recursion relation appropriate for the operation in that item;
3. Organize the subprograms and the data handling so that the *derived program* will accept initial values for N, t_0, $x(t_0)$, (and $x'(t_0)$, $x''(t_0)$, $\cdots$ if required); the derived program will evaluate and store, in order, the first Taylor coefficients of each item in the list, then the second Taylor coefficient of each item in the list (which may require some of the stored values of the first coefficients and initial data) etc., until the entire array of coefficients has been computed; the process can be carried out either in real computer arithmetic or in rounded interval arithmetic;
4. The list in step 1 above can be generated from FORTRAN (or other programming language) subprogram description of $x(t)$ (N 4).

Some examples will help to make the method clear. Consider the function $x(t)$ defined (when t_0 and $x(t_0)$ are given) by the differential equation

(3.20) $x'(t) = x^2 + t^2.$

We have the list

$$T_1 = t \cdot t,$$

(3.21) $$T_2 = (x)_0 \cdot (x)_0,$$

$$(x)_1 = T_1 + T_2.$$

Applying the recursion relations given above, we obtain

$$(T_1)_k = \sum_{j=0}^{k} (t)_j (t)_{k-j},$$

(3.22) $$(T_2)_k = \sum_{j=0}^{k} (x)_j (x)_{k-j},$$

$$(x)_{k+1} = ((T_1)_k + (T_2)_k)/(k+1).$$

Since $(t)_1 = 1$ and $(t)_j = 0$ for $j > 1$, the straightforward application of the recursion relations (3.22) could be modified to recognize that $(T_1)_k = 0$ when $k \geq 3$. Even without any such simplification, the recursion relations (3.22) suffice for the numerical evaluation of any number of Taylor coefficients $(x)_k$, $k = 1, 2, \cdots$, for the function $x(t)$. The total number of arithmetic operations required to find $(x)_k$ for $k = 1, 2, \cdots, N$ not taking into account simplifications possible because of zero terms is (from (3.21) and (3.22)): $(N^2 + N - 1)$ additions, $(N^2 + N)$ multiplications, and $(N-1)$ divisions.

The number of arithmetic operations to compute the kth Taylor coefficient of *any* function expressible as a finite combination of rational, elementary, and composition functions grows *at most linearly with k* once the coefficients of lower orders have been obtained and stored by the process described. Thus, the total number of arithmetic operations to obtain the set of Taylor coefficients of orders $1, 2, \cdots, N$ grows no faster than cN^2 for some constant c independent of N and depending only on the particular function in question.

For a more complicated example, consider

$$(3.23) \qquad x(t) = e^{-\sin t} \log_e (1+t).$$

We can represent $x(t)$ by the list

$$(3.24) \qquad \begin{aligned} T_1 &= \sin t, \\ T_2 &= \log_e (1+t), \\ T_3 &= e^{-T_1}, \\ T_4 &= \cos t, \\ x(t) &= T_2 \cdot T_3. \end{aligned}$$

The auxiliary function $T_4 = \cos t$ was introduced because the derivatives of sine and cosine must be computed together by this approach (since they represent components of solutions of a coupled *second order system* of first order differential equations).

Applying the appropriate recursion relations from (3.18) and (3.19) on a line-by-line basis, we obtain, using $(t)_1 = 1$, $(t)_j = 0$ for $j > 1$,

$$(3.25) \qquad \begin{aligned} (T_1)_k &= (T_4)_{k-1}/k, \\ (T_2)_k &= (a_k - (1-1/k)(T_2)_{k-1})/(1+t), \\ (T_3)_k &= \sum_{j=0}^{k-1} (1-j/k)(T_3)_j(-T_1)_{k-j}, \\ (T_4)_k &= -(T_1)_{k-1}/k, \\ (x)_k &= \sum_{j=0}^{k} (T_2)_j(T_3)_{k-j} \end{aligned}$$

where $a_k = 1$ for $k = 1$ and $a_k = 0$ for $k > 1$. Again the total arithmetic computation required to find the Taylor coefficients of orders $1, 2, \cdots, N$ for $x(t)$ defined by (3.23) at any value of t is of order N^2, (about $\frac{3}{2}N^2$ additions and multiplications). Note that the functions $\sin t$, $\cos t$, $\log_e (1+t)$, $e^{-\sin t}$ need only be evaluated *once* each, at the outset of the process.

We can bound the remainder term in the finite Taylor series (3.15) for this function $x(t)$ (given in (3.23)) by using interval extensions of the functions $\sin t$,

$\cos t$, $\log_e (1+t)$, and e^{-T_1}. Carrying out evaluations of those interval extensions for the interval $[t_0, t]$ and following this with evaluations of the Taylor coefficients, using (3.25) *in interval arithmetic*, we can obtain the quantity $R_N([t_0, t])$ in (3.16).

3.5. Interval polynomial enclosures. Let T_0 be an interval and let x be a real valued function of the real variable t, with $x(t)$ defined for all t in T_0. By an *interval enclosure* of x we mean an inclusion monotonic interval valued function X of an interval variable T, with $X(T)$ defined for all $T \subseteq T_0$ having the property

$$(3.26) \qquad x(t) \in X(t) \quad \text{for all } t \text{ in } T_0.$$

In (3.26) it is not necessary that $X(t)$ be real valued. It may be a nondegenerate interval for each t. If $X(t)$ is a polynomial in t with interval coefficients, it is called an *interval polynomial enclosure*. If $X(t) = x(t)$ for all t in T_0 (see § 3.3) then X is an inclusion monotonic interval extension of x. For an interval encosure X of a real function x on T_0, we have

$$(3.27) \qquad \bar{x}(T) = \{x(t): t \in T\} \subseteq X(T) \quad \text{for all } T \subseteq T_0.$$

Example 1. If $x(t)$ is defined for t in T_0 and $a \leq x(t) \leq b$ for all t in T_0, then the *constant interval function*, $X(T) \equiv [a, b]$, is an interval enclosure of x.

Example 2. If $x(t)$ is continuously differentiable on T_0, and $X'(T)$ is an inclusion monotonic interval extension of $x'(t)$, then, by the mean value theorem, for any t_0 in T_0, we have

$$(3.28) \qquad x(t) = x(t_0) + x'(s)(t - t_0) \quad \text{for some } s \text{ between } t \text{ and } t_0.$$

It follows that $X(T) = x(t_0) + X'(T_0)(T - t_0)$ is an interval enclosure of x. The interval valued function $X(t)$ of the real variable, t, is an interval polynomial enclosure of x. We have

$$(3.29) \qquad x(t) \in X(t) = x(t_0) + X'(T_0)(t - t_0) \quad \text{for all } t \text{ in } T_0.$$

Example 3. If $x(t)$ is analytic for t in T_0, and if the Nth Taylor coefficient of x has an inclusion monotonic interval extension, R_N, then the interval polynomial

$$(3.30) \qquad X_N(t) = \sum_{k=0}^{N-1} (x)_k (t - t_0)^k + R_N(T_0)(t - t_0)^N$$

is an enclosure of x. The values of $X_N(T)$ for nondegenerate intervals $T \subseteq T_0$ are also defined and we have

$$(3.31) \qquad x(t) \in X_N(t) \subseteq X_N(T) = \sum_{k=0}^{N-1} (x)_k (T - t_0)^k + R_N(T_0)(T - t_0)^N$$

for all t in $T \subseteq T_0$.

Example 4. For a specific example of an enclosure of the form (3.30), let $x(t) = e^t$ and take $t_0 = 0$, $T_0 = [0, 1]$, $N = 2$. We have

$$e^t \in 1 + t + ([1, e]/2)t^2 \quad \text{for all } t \text{ in } [0, 1].$$

For a finite representation, which is still an interval polynomial enclosure, we can take $X(t) = 1 + t + [.5, 1.359141]t^2$. For a sharper enclosure, we can take $N = 10$ in (3.30) to obtain

$$e^t \in X_{10}(t) = 1 + t + \tfrac{1}{2}t^2 + \cdots + (1/9!)t^9 + ([1, e]/10!)t^{10}.$$

Only the last coefficient is a nondegenerate interval. It is contained in the finitely represented interval $[.27 \cdot 10^{-6}, .75 \cdot 10^{-6}]$. We can enclose the real coefficients in narrow intervals to obtain a finitely represented interval polynomial enclosure $X(t)$ which contains $X_{10}(t)$ for every t in $[0, 1]$. Of course, $X(t)$ also contains e^t for every t in $[0, 1]$ (N 6).

Example 5. For the sine function, we have the interval polynomial enclosures (for any positive integer N):

$$\sin t \in t - (\tfrac{1}{3}!)t^3 + \cdots + (-1)^{N-1}(1/(2N-1)!)t^{2N-1}$$

$$+ ([-1, 1]/(2N+1)!)t^{2N+1} \quad \text{for any } t.$$

In applications of interval polynomial enclosures to the iterative solution of operator equations (differential and integral equations in particular), sequences of interval polynomials are sometimes constructed with rapidly increasing degrees. An important technique for restraining the unbounded growth of degrees of a sequence of interval polynomials generated by an iterative procedure is due to F. Kruckeberg (N 7), who calls the technique *coarsening* ("Vergröberung").

If m and n are positive integers with $m \leq n$, and if t lies in the unit interval $[0, 1]$ then $0 \leq t^n \leq t^m$. From this observation, it follows that $t^n \in [0, 1]t^m$ whenever $m \leq n$ and $t \in [0, 1]$. Similarly, if t lies in T_0, and $m \leq n$, then $(t - \underline{T}_0)/(\bar{T}_0 - \underline{T}_0)$ lies in $[0, 1]$ and so $(t - \underline{T}_0)^n \in [0, 1](\bar{T}_0 - \underline{T}_0)^{n-m}(t - \underline{T}_0)^m$.

We can also write this last relation in the form

$$(3.32) \qquad (t - \underline{T}_0)^n \in [0, w(T_0)^{n-m}](t - \underline{T}_0)^m \quad \text{for } t \in T_0, m \leq n.$$

Using the relation (3.32), we can *reduce the degree* of an interval polynomial enclosure. By "coarsening", we can obtain a wider polynomial enclosure of lower degree.

Compare the following two interval polynomial enclosures of $\sin t$. Let $X_1(t) = t - (\tfrac{1}{3}!)t^3 + ([-1, 1]/5!)t^5$ and $X_2(t) = t - ((\tfrac{1}{3}!) + [-1, 1]/5!)[0, \bar{T}_0^2])t^3$ for t in $T_0 = [0, \bar{T}_0]$. Both $X_1(t)$ and $X_2(t)$ contain the value of $\sin t$ for each t in $[0, \bar{T}_0]$. While $X_1(t)$ is of degree five, $X_2(t)$ is only of degree three. The widths of the two intervals $X_1(t)$ and $X_2(t)$ for a given t are $w(X_1(t)) = \tfrac{2}{120}t^5$ and $w(X_2(t)) = \tfrac{2}{120}\bar{T}_0^2 t^3$, respectively. The *maximum* width, for t in T_0 is the *same* for both!

NOTES

1. See W. Strother [94], [95]. I am indebted to Wayman Strother for stimulating discussions on fixed points of set valued mappings, continuity and integration of multi-valued functions, and other topics of fundamental importance during the early development of interval analysis.

2. See U. Kulisch [48] and also G. Alefeld and J. Herzberger [1a].

3. See J. M. Yohe [100].

4. See R. E. Moore [57, chap. 11], L. B. Rall [77, § 24]. J. H. Gray and L. B. Rall [19]; D. Kuba and L. B. Rall [47]. The latter two reports describe programs which can automatically obtain values of the partial derivatives in the Jacobian matrix for a system of nonlinear equations, given only the definitions of the functions in the systems. See also R. E. Moore [59, pp. 134–138] and G. Kedem [42].

5. In P. Henrici [33, p. 66], the reader is asked to verify that expressions for the successive derivatives of y defined by $y' = x^2 + y^2$ increase in complexity as the order increases. This is, indeed, the case. Nevertheless, since *we do not need these expressions* in order to *evaluate* derivatives (as is shown in § 3.4 of this book) the remark has *no relevance* as far as criticism of the Taylor series method is concerned. In fact, in R. E. Moore [59, p. 161], it is shown that we can use the recursion relations (for the above differential equation, in the notation $x' = t^2 + x^2$)

$$(x)_1 = t^2 + x^2,$$

$$(x)_2 = t + (x)_0(x)_1,$$

$$(x)_3 = (1 + 2(x)_0(x)_2 + (x)_1^2)/3,$$

$$(x)_4 = ((x)_0(x)_3 + (x)_1(x)_2)/2,$$

to evaluate the Taylor coefficients needed for the fourth order Taylor series method

$$x(t+h) = \sum_{k=0}^{4} (x)_k h^k$$

$$= h(h(h(h(x)_4 + (x)_3) + (x)_2) + (x)_1) + (x)_0.$$

The total number of operations for one fourth order Taylor step for this differential equation is: 9 additions, 14 multiplications. In contrast, the standard fourth order Runge–Kutta formula requires 13 additions and 15 multiplications per step. For this problem, *which was used as an argument against the Taylor series method*, the Taylor series method is *faster* than the widely used Runge–Kutta method *of the same order.*

In 1965, after seeing this approach, P. Henrici, in [76, Vol. I, p. 188], wrote "it used to be part of the folklore of numerical analysis that such derivatives are difficult to evaluate, and this led to the development of increment functions of the Runge–Kutta type, which produce close agreement using judicious combinations of values of the function f alone. However, in the paper by R. E. Moore already referred to, it is shown how, at least for rational functions f, [see also § 3.19 of *this* book] the machine can be used to evaluate the total derivatives of f recursively, *thus removing the main objection to the Taylor expansion method.*" [Italics by author.] Henrici is correct in everything here except in sayng that "it *used to be* part of the folklore". It still is. See almost any recent elementary text on numerical methods. On the other hand, see [59 (especially Appendix C!)].

6. See R. E. Moore [25, p. 7], [58], [62].

7. See F. Krückeberg [46] and also W. Appelt [70, pp. 141–149], and J. Rokne [85]. Chebyshev approximation and "economization" methods improve the accuracy of the polynomial of reduced degree.

Chapter 4

Finite Convergence

4.1.   Interval topology. We introduce a *metric topology* for the set of intervals as follows. We define $d(X, Y) = \max(|\underline{X} - \underline{Y}|, |\bar{X} - \bar{Y}|)$. We call $d(X, Y)$ the *distance* between X and Y. It is easy to see that the defining properties of a metric are satisfied by $d(X, Y)$:

(4.1)
$$d(X, Y) = 0 \quad \text{if and only if } X = Y;$$
$$d(X, Y) = d(Y, X),$$
$$d(X, Z) \leqq d(X, Y) + d(Y, Z).$$

The real line is "isometrically" embedded in our metric space of intervals, since, for degenerate intervals $[x, y]$ and $[y, y]$ we have $d([x, x], [y, y]) = |x - y|$, which is the usual topology on the real line.

We define *continuity* and *uniform continuity* for interval valued functions in the usual ε-δ fashion with the metric $d(X, Y)$. It is not hard to show that the interval arithmetic operations and rational interval functions are continuous as long as no division by an interval containing zero occurs (N 1).

It is useful to note the following. Suppose that $F(X)$ is a natural interval extension of a real rational function $f(x)$; and suppose we can evaluate $F(x_0)$ for some interval X_0 without encountering a division by an interval containing zero. Then the range of values of $f(x)$ for x in X_0 is bounded by $\underline{F(X_0)} \leqq f(x) \leqq \overline{F(X_0)}$ and so $f(x)$ cannot have any poles in X_0. Furthermore, by the inclusion monotonicity of the interval arithmetic operations (and hence of $F(X)$), no division by an interval containing zero will occur during the evaluation of the same expression for $F(X)$ for any $X \subseteq X_0$. In other words, if $F(X_0)$ is defined, then so is $F(X)$ for every $X \subseteq X_0$.

33

We can list some properties of the interval metric which can sometimes be useful:

$$d(X+Z, Y+Z) = d(X, Y), \quad \text{for any interval } Z;$$

(4.2) $X \subseteq Y$ implies $d(X, Y) \leq w(Y) = \bar{Y} - \underline{Y};$

$$d(X, 0) = |X| = \max(|\underline{X}|, |\bar{X}|).$$

We will need the following lemma, which states that natural interval extensions of real rational functions satisfy a sort of Lipschitz condition.

LEMMA 4.1. *If F is a natural interval extension of a real rational function with $F(X)$ defined for $X \subseteq X_0$, where X and X_0 are intervals or n-dimensional interval vectors, then there is a constant L such that $w(F(X)) \leq Lw(X)$ for all $X \subseteq X_0$ (where $w(X) = \max_i w(X_i), i = 1, 2, \cdots, n$).*

Proof. For any real numbers a, b and any intervals X_i, Y_j, we have the following relations (which are not hard to prove):

$$w(aX_i + bY_j) = |a|w(X_i) + |b|w(Y_j),$$

(4.3) $$w(X_iY_j) \leq |X_i|w(Y_j) + |Y_j|w(X_i),$$

$$w(1/Y_j) \leq |1/Y_j|^2 w(Y_j), \quad \text{for } Y_j \text{ not containing } 0.$$

Since the natural interval extension has interval values $F(X)$ obtained by a fixed finite sequence of interval arithmetic operations on real constants (from a given finite set of coefficients) and on the components of X (if X is an interval vector) and since $X \subseteq X_0$ implies that $|X_i| \leq \|X_0\|$ (see (2.6)) for every component of X, it follows that a finite number of applications of the relations (4.3) will produce a constant L such that $w(F(X)) \leq Lw(X)$ for all $X \subseteq X_0$ as desired.

In order to be able to extend results based on Lemma 4.1 to certain interval extensions of irrational functions, we make the following:

DEFINITION. An interval extension F is *Lipschitz* in X_0 if there is a constant L such that $w(F(X)) \leq Lw(X)$ for every $X \subseteq X_0$.

What the condition says is that the width of the image goes to zero at least linearly with the width of the argument, (the argument X may be an interval or an interval vector, $X = (X_1, X_2, \cdots, X_n)$).

LEMMA 4.2. *If a real valued function $f(x)$ satisfies an ordinary Lipschitz condition in $X_0, |f(x) - f(y)| \leq L|x - y|$ for x, y in X_0, then the united extension of f is a Lipschitz interval extension in X_0.*

Proof. The function f is necessarily continuous. The interval (or interval vector) X_0 is compact. Thus, $w(\bar{f}(X)) = |f(x_1) - f(x_2)|$ for some x_1, x_2 in $X \subseteq X_0$. But $|x_1 - x_2| \leq w(X)$; therefore, $w(\bar{f}(X)) \leq Lw(X)$ for $X \subseteq X_0$.

Thus, from Lemma 4.1, a rational interval extension $F(X)$ defined for $X \subseteq X_0$ is Lipschitz in X_0. It follows from Lemma 4.2 that interval extensions which are united extensions, such as the following, are also Lipschitz in X_0: X^n

given by (3.10); $e^X = [e^{\underline{X}}, e^{\bar{X}}]$; $X^{1/2} = [\underline{X}^{1/2}, \bar{X}^{1/2}]$ for $0 < \underline{X}_0$; $\log_e X = [\log_e \underline{X}, \log_e \bar{X}]$ for $0 < \underline{X}_0$; $\sin X = [\sin \underline{X}, \sin \bar{X}]$ for $X_0 \subseteq [-\pi/2, +\pi/2]$, $\sin X = [\min_{x \in X} (\sin x), \max_{x \in X} (\sin x)]$ for arbitrary X_0.

If F and G are inclusion monotonic interval extensions with F Lipschitz in Y_0, G Lipschitz in X_0 and $G(X_0) \subseteq Y_0$, then the composition $H(X) = F(G(X))$ is Lipschitz in X_0 and is inclusion monotonic. Thus, $\sin G(X)$ is Lipschitz and inclusion monotonic for X in X_0 if $G(X)$ is rational in X and $G(X_0)$ is defined. Ordinary monotonicity is *not* required for inclusion monotonicity.

Thus, an inclusion monotonic interval extension whose values $F(X)$ are defined by a fixed finite sequence of rational (interval arithmetic) operations and/or compositions of Lipschitz extensions is, itself, a Lipschitz extension in some suitably chosen region.

If X and Y are intervals such that $X \subseteq Y$, then there is an interval E with $\underline{E} \leq 0 \leq \bar{E}$ such that $Y = X + E$ and $w(Y) = w(X) + w(E)$. If F is an inclusion monotonic interval extension of f with $F(X)$ defined for $X \subseteq X_0$, then $\bar{f}(X) \subseteq F(X)$ for $X \subseteq X_0$. We have $F(X) = \bar{f}(X) + E(X)$ for some interval valued function $E(X)$ with $w(F(X)) = w(\bar{f}(X)) + w(E(X))$. We call $w(E(X)) = w(F(X)) - w(\bar{f}(X))$ the *excess width* of $F(X)$.

The united extension itself has, of course, *zero* excess width. Its interval values $\bar{f}(X)$ give *exactly* the range of values of a continuous function $f(x)$ for x in X.

By a *uniform subdivision* of an interval vector $X = (X_1, X_2, \cdots, X_n)$ we mean the following. Let N be a positive integer. Define $X_{i,j} = [\underline{X}_i + (j-1)w(X_i)/N, \underline{X}_i + jw(X_i)/N]$, $j = 1, 2, \cdots, N$. We have $X_i = \bigcup_{j=1}^{N} X_{i,j}$ and $w(X_{i,j}) = w(X_i)/N$. Furthermore, $X = \bigcup_{j_i=1}^{N} (X_{1,j_1}, X_{2,j_2}, \cdots, X_{n,j_n})$ with $w(X_{1,j_1}, \cdots, X_{n,j_n}) = w(X)/N$.

We have the following important result.

THEOREM 4.1. *If $F(X)$ is an inclusion monotonic, Lipschitz, interval extension for $X \subseteq X_0$, then the excess width of a refinement, $F_{(N)}(X)$, the union of interval values of F on the elements of a uniform subdivision of X, is of order $1/N$. We have*

$$(4.4) \qquad F_{(N)}(X) = \bigcup_{j_i=1}^{N} F(X_{1,j_1}, \cdots, X_{n,j_n}) = \bar{f}(X_1, \cdots, X_n) + E_N$$

and there is a constant K such that

$$(4.5) \qquad w(E_N) \leq Kw(X)/N.$$

Proof. The relation (4.4) follows from the fact that $\bar{f}(X) = \bigcup_s \bar{f}(X_s)$ where the X_s are the elements of the subdivision. We have $F(X_s) = \bar{f}(X_s) + E_s$ for some E_s and

$$w(E_s) = w(F(X_s)) - w(\bar{f}(X_s)) \leq w(F(X_s)) \leq Lw(X_s) \leq Lw(X)/N$$

for every X_s in the uniform subdivision. Clearly, $w(E_N) \leq 2Lw(X)/N$, yielding

(4.5), with $K = 2L$, (N 2); since, in the worst case, the maximum excess width may have to be added to both upper *and* lower bounds in the union.

From Theorem 4.1 it follows that we can compute arbitrarily sharp upper and lower bounds on the exact range of values of a wide variety of real valued functions of n variables by subdividing the domain of arguments and taking the union of interval evaluations over the elements of the subdivision. In the last section of this chapter we will discuss some more efficient means for doing the same thing.

4.2.   Finite convergence.   A sequence of intervals, $\{X_k\}$, is said to be *nested* if $X_{k+1} \subseteq X_k$ for all k. A sequence of intervals, $\{X_k\}$, is said to be *convergent* if there is an interval X such that $\lim_{k \to \infty} d(X_k, X) = 0$. Such an interval X is unique and is called the *limit* of the sequence.

LEMMA 4.3. *Every nested sequence of intervals is convergent and has the limit* $X = \bigcap_{k=1}^{\infty} X_k$.

Proof. $\{\underline{X}_k\}$ is a monotone nondecreasing sequence of real numbers, bounded above by $\bar{X}_1$, and so has a limit $\underline{X}$. Similarly, $\{\bar{X}_k\}$ is a monotone nonincreasing sequence of real numbers, bounded below by $\underline{X}_1$, and so has a limit $\bar{X}$. Furthermore, since $\underline{X}_k \leq \bar{X}_k$ for every k, we have $\underline{X} \leq \bar{X}$. Thus $\{X_k\}$ is convergent and $X = [\underline{X}, \bar{X}] = \bigcap_{k=1}^{\infty} X_k$ is its limit.

LEMMA 4.4. *Suppose that X_k is a sequence of intervals such that there is a real number x contained in every X_k. Define the sequence of intervals $\{Y_k\}$ by $Y_1 = X_1$ and $Y_{k+1} = X_{k+1} \cap Y_k$ for $k = 1, 2. \cdots$. Then Y_k is a nested sequence with limit Y and we have*

(4.6) $x \in Y \subseteq Y_k$ *for all k.*

Proof. By induction, the intersection in the definition of Y_{k+1} is nonempty so the sequence $\{Y_k\}$ is well defined. It is nested by construction. The relation (4.6) follows from Lemma 4.3.

By the *finite convergence* of a sequence of intervals $\{X_k\}$, we mean that there is a positive integer K such that $X_k = X_K$ for $k \geq K$. Such a sequence is said to *converge in K steps*. We will now illustrate finite convergence with some examples (N 3).

It is not hard to see that $X_0 = [1, 2]$, $X_{k+1} = 1 + X_k/3$, $k = 0, 1, 2, \cdots$, generates a nested sequence of intervals, $\{X_k\}$. The interval function $F(X) = 1 + X/3$ is inclusion monotonic since it is a rational interval function. Therefore, we have $X_1 = F(X_0) = 1 + [1, 2]/3 = [\frac{4}{3}, \frac{5}{3}] \subseteq X_0 = [1, 2]$. It follows that $X_{k+1} = F(X_k) \subseteq X_k$ for all k, by finite induction. By Lemma 4.1, the sequence has a limit, X. If we compute the elements of the sequence $\{X_k\}$ using rounded interval arithmetic, then we will obtain another sequence $\{X_k^*\}$ with $X_k \subseteq X_k^*$ for all k. More precisely, let the sequence X_k^* be defined by $X_0^* = X_0 = [1, 2]$ and $X_{k+1}^* = \{1 + X_k^*/3 : \text{computed in rounded interval arithmetic}\} \cap X_k^*$, $k = 0, 1, 2, \cdots$, (which we can apply, since $X \subseteq X_k$ for all k). It follows from

Lemma 4.4 that X_k^* is a nested sequence and that the limit of the sequence $\{X_k\}$ is contained in the limit, X^*, of the sequence $\{X_k^*\}$. The sequence $\{X_k^*\}$ will converge in a finite number of steps. For instance, if we use three decimal digit rounded interval arithmetic, we find, for the sequence $\{X_k^*\}$,

$$X_0^* = [1, 2],$$

$$X_1^* = [1.33, 1.67],$$

$$X_2^* = [1.44, 1.56],$$

$$X_3^* = [1.48, 1.52],$$

$$X_4^* = [1.49, 1.51],$$

$$X_5^* = [1.49, 1.51],$$

and $X_k^* = X^*$ for all $k \geq 4$. We have finite convergence in four steps. Note that the real sequence $x_{k+1} = 1 + x_k/3$ converges to 1.50 *in the limit* (after *infinitely* many steps) from any x_0.

As a second example of finite convergence we consider a sequence of refinements computed in rounded interval arithmetic and intersected to yield a nested sequence of intervals which converges in a finite number of steps. Consider the interval polynomial $F(X) = X(1 - X)$ and its refinements $F_{(N)}(X)$ defined by (4.4). Let $X = [0, 1]$ and consider the sequence of intervals $Y_1 = F_{(1)}([0, 1]) = F([0, 1])$, $Y_{k+1} = F_{(k+1)}([0, 1]) \cap Y_k$, $k = 1, 2, \cdots$. The intersections are nonempty because each refinement contains the range of values of $f(x) = x(1 - x)$ for x in $[0, 1]$, (namely $[0, \frac{1}{4}]$). By construction, $\{Y_k\}$ is a nested sequence. Thus we have, by Theorem 4.1 and Lemma 4.3, the result

$$(4.7) \qquad \bar{f}(X) = \bigcap_{k=1}^{\infty} F_{(k)}(X) = \lim_{k \to \infty} Y_k.$$

For k *odd*, we will have

$$F_{(k)}([0, 1]) = [0, \tfrac{1}{4} + 1/(2k) + 1/(4k^2)];$$

and for k *even*, we will have

$$F_{(k)}([0, 1]) = [0, \tfrac{1}{4} + 1/(2k)].$$

We can compute a *subsequence* of the sequence $\{Y_k\}$, for example for k of the form $k = 2^t$, $t = 1, 2, \cdots$. If we compute, then, the nested sequence of intervals $Y_1^* = [0, 1]$, and for $t = 1, 2, \cdots$,

$$Y_{t+1}^* = F_{(2^t)}^*([0, 1]) \cap Y_t^*$$

where $F_{(k)}^*([0, 1])$ is $F_{(k)}([0, 1])$ computed in rounded interval arithmetic to three decimal digits, we obtain again a nested sequence $\{Y_t^*\}$. This sequence converges in nine steps to the interval $[0, .251]$. We have $Y_t^* = [0, .251]$ for all $t \geq 9$.

For any fixed precision representation of machine numbers (as opposed to "variable" precision), there is a *finite set* of machine numbers represented by strings of bits $b_0 b_1, \cdots, b_s$ with s *fixed*; in this case, there is only a finite set of intervals with machine number end points. *Any nested sequence of such intervals is necessarily finitely convergent.*

4.3. Stopping criteria. For any iterative interval method which produces a *nested sequence* of intervals whose end points are represented by fixed precision machine numbers, we have a natural stopping criterion. Since the sequence will converge in a finite number of steps, we can compute the elements X_k of the sequence $\{X_k\}$ until we reach the condition $X_{k+1} = X_k$. If the elements X_k are generated by an iteration procedure of the form

$$(4.8) \qquad\qquad X_{k+1} = F(X_k)$$

such that each X_{k+1} depends only on the previous X_k, then, clearly, the criterion $X_{k+1} = X_k$ will, if satisfied, guarantee that the sequence has converged.

In particular, if $F(X)$ is a rational expression in X and if X_0 is an interval such that $F(X_0) \subseteq X_0$, *which can be tested on the computer*, then it follows that the sequence of intervals defined by

$$X_{k+1} = F(X_k), \qquad k = 0, 1, 2, \cdots,$$

is a *nested* sequence

$$X_0 \supseteq X_1 \supseteq X_2 \supseteq \cdots$$

and hence *converges* to some interval

$$X^* \quad \text{with } X^* = F(X^*) \text{ and } X^* \subseteq X_k$$

for all $k = 0, 1, 2, \cdots$.

On the computer, rounding right end points to the right and left end points to the left when necessary during rounded interval arithmetic, it may happen that

$$X_1 = F(X_0) \subseteq X_0$$

but that for some k, X_{k+1} is not contained in X_k.

If we modify the straightforward iteration procedure to compute instead

$$X_{k+1} = F(X_k) \cap X_k$$

and stop when $X_{k+1} = X_k$ (which will happen always for some k on a computer using fixed finite precision interval arithmetic properly rounded) then the computer will have obtained the narrowest interval possible containing $X^* = F(X^*)$. A narrower interval would require using higher precision arithmetic.

Furthermore, if for a chosen X_0, the interval $F(X_0) \cap X_0$ is *empty*, which can also be tested on the computer, then X_0 contains no fixed points of F, (i.e. there is no X in X_0 such that $F(X) = X$). This follows from the inclusion mono-tonicity of F; for, if $F(X) = X$ and $X \subseteq X_0$ then $X = F(X) \subseteq F(X_0)$ and so $X \subseteq F(X_0) \cap X_0$; therefore, if $F(X_0) \cap X_0$ is empty, there is no such X.

A simple example will illustrate some of the things that can happen. Let

$$F(X) = \tfrac{1}{2}X + 2.$$

If we take $X_0 = [1, 2]$, then

$$F(X_0) = \tfrac{1}{2}[1, 2] + 2 = [\tfrac{5}{2}, 3].$$

In this case $F(X_0)$ does not intersect X_0; that is, $F(X_0) \cap X_0$ is empty and so there is no fixed point of F in $[1, 2]$.

If we take, instead, $X_0 = [2, \tfrac{7}{2}]$, then

$$F(X_0) = F([2, \tfrac{7}{2}]) = \tfrac{1}{2}[2, \tfrac{7}{2}] + 2$$

$$= [1, \tfrac{7}{4}] + 2$$

$$= [3, \tfrac{15}{4}].$$

In this instance $F(X_0)$ and X_0 do intersect; in fact $F(X_0) \cap X_0 = [3, \tfrac{7}{2}]$. We cannot conclude anything in this case since $F(X_0)$ is not *contained* in X_0.

Suppose we take $X_0 = [2, 5]$; then $F(X_0) = F([2, 5]) = \tfrac{1}{2}[2, 5] + 2 = [3, \tfrac{9}{2}]$. In this case $F(X_0) \subseteq X_0$ and so F has a fixed point X in X_0. The iterations

(4.8)′ $$X_{k+1} = F(X_k) \cap X_k, \qquad k = 0, 1, 2, \cdots,$$

produce, in *properly rounded three decimal digit interval arithmetic*:

$$X_1 = [3, 4.5] \cap [2, 5] = [3, 4.5],$$

$$X_2 = [3.5, 4.25] \cap [3, 4.5] = [3.5, 4.25],$$

$$X_3 = [3.75, 4.13] \cap [3.5, 4.25] = [3.75, 4.13],$$

$$X_4 = [3.87, 4.07],$$

$$X_5 = [3.93, 4.04],$$

$$X_6 = [3.96, 4.02],$$

$$X_7 = [3.98, 4.01],$$

$$X_8 = [3.99, 4.01],$$

$$X_9 = [3.99, 4.01],$$

$$X_{k+1} = X_k, \qquad k = 8, 9, 10, \cdots.$$

There is a fixed point of F in $[3.99, 4.01]$.

If the process generating the sequence depends *explicitly* on k as well as on X_k, (say $X_{k+1} = F(k, X_k)$), then we might have $X_{k+1} = X_k$ for some k and yet $X_{k+2} \neq X_k$, even though $\{X_k\}$ is a nested sequence. An example is

(4.9) $X_{k+1} = ([0, 2]/k) \cap X_k, \qquad X_1 = [0, 1].$

We have

$$X_1 = [0, 1],$$
$$X_2 = [0, 1],$$
$$X_3 = [0, 1],$$
$$X_4 = [0, \tfrac{2}{3}],$$
$$X_5 = [0, \tfrac{1}{2}],$$
$$\cdots$$
$$X_{k+1} = [0, 2/k], \qquad k > 2.$$

Thus, $X_{k+1} = X_k$ is a valid stopping criterion if and only if the sequence $\{X_k\}$ is nested and is generated by (4.8)' with $F(X_k)$ depending *only* on the value of X_k.

For real numbers x and y, the relation $x \leq y$ is either *true* or *false*. If it is false, then we have $x > y$. If all we know about x and y is that they lie in intervals $x \in X$ and $y \in Y$, then we can only deduce one of *three* possibilities:

(1)   If $\bar{X} \leq \underline{Y}$, then $x \leq y$,

(4.10)   (2)   If $\underline{X} > \bar{Y}$, then $x > y$,

(3)   otherwise, we *don't know* whether $x \leq y$ or $x > y$.

We can modify any computer program by using rounded interval arithmetic instead of ordinary machine arithmetic whenever rounding error is possible in an arithmetic operation and by using the above *three-valued logic* for all branch tests involving computed quantities which are only known to lie in certain computed intervals. If we *stop* the computation whenever a "don't know" logical value is obtained, then all intervals computed up to that point by the modified problem will contain the exact result of the corresponding finite sequence of operations using exact (infinite precision) real arithmetic for all arithmetic operations (N 4).

4.4.   More efficient refinements. In § 4.1, we introduced the refinement of an interval extension as a method for computing arbitrarily sharp upper and lower bounds on the range of values of a real function. For applications of interval methods it is of practical importance to be able to obtain good bounds on the ranges of values of real functions as efficiently as possible. In this section we will explore some ways of improving the efficiency of the straightforward refinement given by (4.4).

We will begin with a characterization of the class of *functions most commonly used in computing.* Let $FC_n(X_0)$ be the class of real valued functions f of n real variables whose values $f(x_1, x_2, \cdots, x_n)$ for each f are defined for all $x = (x_1, x_2, \cdots, x_n)$ in the interval vector $X_0 = (X_{10}, X_{20}, \cdots, X_{n0})$ by some fixed, finite sequence of operations

$$T_1 = h_1(y_1, z_1),$$

(4.11) $$T_2 = h_2(y_2, z_2),$$

$$\cdots$$

$$f(x) = T_M = h_M(y_M, z_M)$$

for $y_1, z_1 \in S_1 = \{x_1, x_2, \cdots, x_n, c_1, c_2, \cdots, c_p\}$, where $c_1, c_2, \cdots, c_p$ are given real numbers (called the "coefficients" of f), and where $h_1, h_2, \cdots, h_M$ are the arithmetic functions $+, -, \cdot, /,$ or *unary* functions (of only *one* of the variables y_i, z_i) of elementary type: $\exp, \log_e, \sqrt{\cdot},$ etc. Furthermore, $y_{i+1}, z_{i+1} \in S_{i+1} = S_i \cup \{T_i\},$ $i = 1, 2, \cdots, M-1$. The integer M, the coefficients $c_1, \cdots, c_p$ (if any), and the sequence of operations $h_1, h_2, \cdots, h_M$ define a particular function $f \in FC_n(X_0)$. Each y_i, z_i may be an x; or a constant c_j or one of the previous T_j ($j < i$).

We will assume here that we can compute arbitrarily sharp bounds on the exact range of values of each of the unary functions (see (3.10) and the discussion following it) occurring in the sequence of operations for $f(x)$. The class of functions described, FC_n, is further assumed to include only those functions f which are defined for all x in X_0, and which have inclusion monotonic, Lipschitz, interval extensions $F(X)$ for $X \subseteq X_0$. For some of the discussion to follow, we will also assume that the first and second partial derivatives of f satisfy the same conditions as f. This final assumption, concerning the existence of interval extensions with desirable properties (inclusion monotonicity, etc.), *will be satisfied automatically* for *all* functions f in $FC_n(X_0)$, provided only that $F(X_0)$ is defined, where $F(X)$ is the interval extension

$$T_1 = H_1(Y_1, Z_1),$$

(4.12) $$T_2 = H_2(Y_2, Z_2),$$

$$\cdots$$

$$F(X) = T_M = H_M(Y_M, Z_M)$$

where $H_1, H_2, \cdots, H_M$ are the united extensions of the functions $h_1, h_2, \cdots, h_M$ respectively. This will be the case, replacing real arithmetic operations by interval arithmetic operations and using the united extensions of the unary functions, as long as for the interval vector X_0, no division by an interval containing 0 occurs during the computation of $T_1, T_2, \cdots, T_M$ and provided that no arguments containing zero occur for such unary operations as $\log_e$ or $\sqrt{\cdot}$. (The latter would spoil the Lipschitz property for the interval extension.) Since *such an F is inclusion monotonic*, $F(X)$ is defined for every $X \subseteq X_0$.

Under these conditions, we can apply Theorem 4.1 and find that the excess width of the refinement $F_{(N)}(X)$ of the interval extension $F(X)$ given by (4.12) of a function f in $FC_n(X_0)$ satisfies

$$(4.13) \qquad w(E_N) \leqq K_f w(X)/N \qquad \text{for some } K_f \text{ depending only on } f.$$

An evaluation of F requires one pass through the finite sequence of operations (4.12); just as an evaluation of f requires one pass through the finite sequence (4.11). An evaluation of $F_{(N)}(X)$, given by (4.4) requires N^n evaluations of $F(X)$. If n is at all large, this would involve a prohibitive amount of computation to achieve the result (4.13) for large N. Even for $n = 2$ we have, for $N = 1000$ (perhaps to reduce the excess width to .001), $1000^2 = 10^6$ evaluations to carry out.

Fortunately, *by using more information* about a particular function f, we can compute an interval, containing the exact range of values, of arbitrarily small excess width with far less work. In some cases we can even compute an interval with zero excess width in *one* evaluation of $F(X)$. In the remainder of this section the function f will be assumed to lie in $FC_n(X_0)$ and $F(X)$ is defined for $X \subseteq X_0$ by (4.12).

THEOREM 4.2. *If each of the variables $x_1, x_2, \cdots, x_n$ occurs at most once in the list $y_1, z_1, y_2, z_2, \cdots, y_n, z_n$ of arguments in the sequence (4.11) defining $f(x)$, then the interval extension $F(X)$ defined by (4.12) is the united extension of f for all $X \subseteq X_0$.*

Proof. We will have $F(X) = \bar{f}(X)$ if and only if $\bar{f}(X) \subseteq F(X)$ *and* $F(X) \subseteq \bar{f}(X)$. The first inclusion is always true, under the assumptions we have made on $F(X)$. Under the further hypothesis of this theorem, every real number r, in the interval $F(X)$, is expressible as $r = f(x)$ for some $x = (x_1, x_2, \cdots, x_n)$ with $x_i \in X_i$ for all $i = 1, 2, \cdots, n$; in fact, each H_i is the united extension of h_i for $i = 1, 2, \cdots, n$. Therefore $F(X) \subseteq \bar{f}(X)$ and the theorem is proved.

The importance of Theorem 4.2 is this: for a function f satisfying the hypotheses of Theorem 4.2, we can compute the exact range of values of $f(x)$ for $x \in X \subseteq X_0$ with *one* evaluation of $F(X)$ using (4.12).

If some x_i occurs more than once in the list of arguments in (4.11), then the corresponding X_i will occur more than once in the list of arguments in (4.12). In this case, there may be real numbers in $F(X)$ which are not expressible as $f(x)$ for any $x \in X$. As an example, consider f in $FC_1([0, 1])$ defined by

$$T_1 = 1 - x_1, \qquad f(x_1) = T_2 = T_1 x_1.$$

Here, the variable x_1 occurs twice as an argument. Consider the corresponding list for $F(X)$. We have

$$T_1 = 1 - X_1, \qquad F(X_1) = T_2 = T_1 X_1.$$

Now $F([0, 1]) = [0, 1]$; but there is *no* x_1 in $[0, 1]$ such that $f(x_1) = (1 - x_1)x_1 = 1 \in [0, 1]$.

On the other hand, the interval function

$$T_1 = 1 - X_1, \qquad F(X_1, X_2) = T_2 = T_1X_2$$

is the united extension of the real function

$$T_1 = 1 - x_1, \qquad f(x_1, x_2) = T_2 = T_1x_2.$$

In particular, $F([0, 1], [0, 1]) = [0, 1]$ is the exact range of values of $f(x_1, x_2) = (1 - x_1)x_2$ for $x_1 \in [0, 1]$ and $x_2 \in [0, 1]$.

Another explanation for the difference between these two results is the following. In both cases we compute, first, $1 - [0, 1] = [0, 1]$. In both cases this is the exact range of values of $1 - x_1$ for x_1 in $[0, 1]$. The discrepancy comes at the next step. When we compute T_1X_1 with $T_1 = [0, 1]$ and $X_1 = [0, 1]$, we obtain $[0, 1]$. We obtain the same result for T_1X_2 with $T_1 = [0, 1]$ and $X_2 = [0, 1]$.

In the second case, this is exactly what we want, the exact range of values of $f(x_1, x_2)$. In the first case, we get an overestimation of the range of values of $f(x_1) = (1 - x_1)x_1$ because, during the evaluation of T_1X_1 numerically, we have not made use of the information that $T_1 = [0, 1]$ represents the range of values of $1 - x_1$ for x_1 in X_1. We have not made any distinction between this and $T_1 = [0, 1]$ as the range of values of $1 - x_2$ for a variable x_2 *independent* of x_1.

Thus, for example, it is better to use the interval extension X^n defined by (3.10) for integer powers x^n than to use $X \cdot X \cdots X$. For $X = [-1, 2]$ we obtain $X \cdot X = [-2, 4]$ whereas X^2 using (3.10) yields the correct result, $[-1, 2]^2 = [0, 4] = \{x^2 : x \in [-1, 2]\}$.

Sometimes we can reduce the number of occurrences of a variable in a real rational expression by algebraic manipulation. For example, let $f(x_1, x_2, x_3)$ be defined by

$$(4.14) \qquad f(x_1, x_2, x_3) = ((x_1 + x_2)/(x_1 - x_2))x_3.$$

We can rewrite this as

$$(4.15) \qquad f(x_1, x_2, x_3) = x_3(1 + 2/((x_1/x_2) - 1)).$$

For $x_1 \in [1, 2]$, $x_2 \in [5, 10]$, $x_3 \in [2, 3]$, the natural interval extension of (4.15) produces the correct range of values since each variable occurs only once. We have

$$[2, 3](1 + 2/(([1, 2]/[5, 10]) - 1)) = [-7, -\tfrac{22}{9}].$$

On the other hand, the natural interval extension of the form (4.14) (which is equivalent in real, but not in interval arithmetic) produces

$$(([1, 2] + [5, 10])/([1, 2] - [5, 10]))[2, 3] = [-12, -\tfrac{12}{9}]$$

with an excess width of $\tfrac{55}{9}$.

Given a function f in the class $FC_n(X_0)$ defined by a finite sequence of the form (4.11), we can construct interval extensions of f other than the natural extension given by (4.12). We will now discuss three such extensions: the *centered form*, the *mean value form*, and the *monotonicity-test form*.

A "centered form" is a particular form of interval extension, $F_c(X_1, \cdots, X_n)$ of a *rational* function $f(x_1, \cdots, x_n)$. To derive F_c for a particular f, we first rewrite $f(x_1, \cdots, x_n)$ as

(4.16) $$f(x_1, \cdots, x_n) = f(c_1, \cdots, c_n) + g(y_1, \cdots, y_n)$$

with $y_i = x_i - c_i$ and with g *defined* by (4.16); thus

$$g(y_1, \cdots, y_n) = f(y_1 + c_1, \cdots, y_n + c_n) - f(c_1, \cdots, c_n).$$

For rational f, it will be possible to write g in the form $g(y_1, \cdots, y_n) = y_1 h_1(y_1, \cdots, y_n) + \cdots + y_n h_n(y_1, \cdots, y_n)$ where h_i is rational and $h_i(0, \cdots, 0)$ is defined provided that $f(c_1, \cdots, c_n)$ is defined. We define F_c by

(4.17) $$F_c(X_1, \cdots, X_n) = f(c_1, \cdots, c_n) + \sum_{i=1}^{n} Y_i H_i(Y_1, \cdots, Y_n)$$

where $c_i = m(X_i)$ and $Y_i = X_i - c_i$ and H_i is the natural interval extension of h_i. An example will illustrate the idea.

We reconsider the polynomial $p(x) = 1 - 5x + \frac{1}{3}x^3$ given before in (3.7). We first rewrite $p(x)$ as

$$p(x) = 1 - 5c + \frac{1}{3}c^3 + g(y) \quad \text{where } y = x - c,$$

and

$$g(y) = p(x) - p(c) = y(-5 + \frac{1}{3}((y + c)^2 + (y + c)c + c^2)).$$

We define P_c by

$$P_c(X) = p(c) + YH(Y)$$

where $c = m(X)$, $Y = X - m(X)$, and $H(Y) = -5 + \frac{1}{3}((Y + c)^2 + (Y + c)c + c^2)$. Now, for $X = [2, 3]$ we obtain: $c = m(X) = \frac{5}{2}$, $Y = [-\frac{1}{2}, \frac{1}{2}]$, and $p(c) = p(\frac{5}{2}) = -\frac{151}{24}$. We have $H(Y) = H([-\frac{1}{2}, \frac{1}{2}]) = [\frac{1}{12}, \frac{31}{12}]$ and so $P_c([2, 3]) = [-\frac{91}{12}, -5] = [-7.5833333\ldots, -5]$. This is an improvement over the bounds computed using the interval extensions (3.8) and (3.9). Recall that the exact range of values of this $p(x)$ for x in $[2, 3]$ is $[-\frac{10}{3}\sqrt{5} + 1, -5] = [-6.454559\ldots, -5]$.

From (4.13) it follows (with $N = 1$) that the excess width of the natural interval extension of a rational function is of order $w(X)$. It has been proved by E. Hansen (N 5) that the excess width of the centered form extension is of order $w(X)^2$.

Thus, if we use the centered form in the computation of the refinements

$F_{(N)}(X)$ defined in (4.4) we will obtain the result

(4.18) $$w(E_N) \leq K_f w(X)/N^2 \quad \text{for some } K_f$$

for the excess width, instead of (4.13). Using the centered form instead of the natural interval extension for rational functions, we could reduce the excess width of refinements to a given amount with about the square root of the number of function evaluations required for the natural extensions. But this is still prohibitively large in many cases and we will discuss ways to make even more dramatic reductions. Nonetheless, for a single evaluation of an interval extension on an interval vector of small width, the centered form can give very small excess width compared to the natural extension for rational functions. Recently, H. Ratschek (N 5) has found higher order centered forms.

Another interval extension of practical value is the *mean value form*. We can apply this technique to functions in the class $FC_n(X_0)$. Let X be an interval vector in X_0 and let $m = m(X)$, (see (2.7)). Let D_iF be an interval extension of $\partial f/\partial x_i$; we will show how to obtain these presently. From the mean value theorem, we have, for all $X \subseteq X_0$,

(4.19) $$\bar{f}(X) \subseteq F_{\text{MV}}(X) = f(m) + \sum_{i=1}^{n} D_iF(X)(X_i - m_i).$$

We call $F_{\text{MV}}(X)$ the *mean value extension* of f on X.

We can illustrate this with an example. Let $f(x_1, x_2)$ be defined by the sequence

$$T_1 = x_2^2,$$

$$T_2 = x_1 + T_1,$$

(4.20) $$T_3 = e^{T_2},$$

$$T_4 = T_3 x_1,$$

$$f(x) = T_4 - T_1.$$

We could also represent f with the expression

$$f(x_1, x_2) = x_1 e^{(x_1 + x_2^2)} - x_2^2.$$

On a line by line basis (see also § 3.4) we can derive D_iF for $i = 1, 2$ as follows. We can use $D_iX_j = 1$ if $j = i$ and 0 if $j \neq i$.

(4.21)

$D_1T_1 = 0,$	$D_2T_1 = 2X_2,$
$D_1T_2 = 1,$	$D_2T_2 = D_2T_1,$
$D_1T_3 = T_3,$	$D_2T_3 = T_3(D_2T_2),$
$D_1T_4 = T_3 + X_1(D_1T_3),$	$D_2T_4 = (D_2T_3)X_1$
$D_1F(X) = D_1T_4,$	$D_2F(X) = D_2T_4 - D_2T_1.$

For $X_1 = [1, 2]$ and $X_2 = [0, 1]$, we find $m = (\frac{3}{2}, \frac{1}{2})$ and, from (4.20), we find $f(m) = (3/2) e^{(7/4)} - \frac{1}{4} = 8.3819040 \ldots$. In the sequence (4.21), the quantities T_1, T_2, T_3, T_4 must first be computed in interval arithmetic using the natural extensions (or united extensions) of the operations in (4.20) beginning with the given interval values for X_1 and X_2. We find

$$T_1 = [0, 1],$$

$$T_2 = [1, 3],$$

$$T_3 = [e^1, e^3] = [2.7182818 \ldots, 20.0855369 \ldots],$$

$$T_4 \subseteq [2.7182818, 40.171074].$$

We next compute

$$D_1 T_1 = 0,$$

$$D_1 T_2 = 1,$$

$$D_1 T_3 = [2.7182818 \ldots, 20.0855369 \ldots],$$

$$D_1 T_4 \subseteq [5.4365636 \ldots, 60.256611],$$

$$D_1 F(X) \subseteq [5.4365636, 60.256611],$$

$$D_2 T_1 = [0, 2],$$

$$D_2 T_2 = [0, 2],$$

$$D_2 T_3 \subseteq [0, 40.171074],$$

$$D_2 T_4 \subseteq [0, 80.342148],$$

$$D_2 F(X) \subseteq [-2, 80.342148].$$

Finally, we compute $F_{MV}([1, 2], [0, 1])$ from (4.19). We obtain

$$F_{MV}([1, 2], [0, 1]) \subseteq [-61.917477, 78.681284].$$

We can obtain narrower bounds on the range of values for many functions in $FC_n(X_0)$ than are provided by the mean value form by using a modification of it which we will call a *monotonicity test form*. A reconsideration of the above example will illustrate the idea. From the fact that $D_1 F(X)$ turned out to lie completely in an interval of positive real numbers, we can conclude that $f(x_1, x_2)$ is monotonic increasing with x_1 for all (x_1, x_2) in $X = ([1, 2], [0, 1])$. Thus, a lower bound on the range of values of $f(x_1, x_2)$ in X can be computed by finding a lower bound on the range of values of $f(1, x_2)$. Similarly, an upper bound can be found by finding an upper bound on the range of values of $f(2, x_2)$. We can do this, for instance with

$$(4.22) \qquad \bar{f}(X) \subseteq [\underline{F_{MV}(1, X_2)}, \overline{F_{MV}(2, X_2)}] \subseteq [-36.9308, 58.8966].$$

For the particular example at hand, the natural extension gives the still sharper bounds [1.7182818, 40.171074].

Now, from (4.21) we have $D_2T_1 = 2X_2, \cdots, D_2F(X) = D_2T_4 - D_2T_1$. If we substitute each expression into the following ones as appropriate, we can rewrite $D_2F(X)$ as

(4.23) $$D_2F(X) = 2X_2(e^{T_2}X_1 - 1),$$

by factoring out $2X_2$. Using (4.23), we find that

(4.24)
$$D_2F([1, 2], [0, 1]) = 2[0, 1](e^{[1,2]+[0,1]^2}[1, 2] - 1)$$
$$= [0, 78.342148].$$

From (4.24) we can conclude that $f(x_1, x_2)$ is also monotonic increasing with x_2 for all (x_1, x_2) in X and so the *exact* range of values can be bounded sharply by computing

(4.25)
$$\bar{f}(X) = [F_{MV}(1, 0), F_{MV}(2, 1)]$$
$$= [f(1, 0), f(2, 1)]$$
$$\subseteq [2.7182818, 39.171074].$$

It follows from a theorem of Alefeld and Herzberger (N 6) that the excess width of the mean value extension is of order $w(X)^2$, just as it is for the centered form. Thus, for interval (or interval vector) arguments which are very narrow (small $x(X)$) both the centered form and the mean value form will give upper and lower bounds which are very close to the exact range of values. For wide intervals (large $w(X)$) the centered or mean value forms may actually give wider bounding intervals than the natural extension—as was the case in the example just discussed. However, for the same function (4.20), but with the narrower argument intervals $X_1 = [1, 1.01]$, $X_2 = [.4, .401]$ we obtain, from (4.19) and (4.21),

$$F_{MV}(X) = [3.029529, 3.096156];$$

whereas we get the wider interval

$$F(x) = [3.029132, 3.096821]$$

using the natural extension

$$F(X) = X_1 e^{X_1 + X_2^3} - X_2^2.$$

The excess width of $F(X)$ here turns out to be .001602 while the excess width of $F_{MV}(X)$ is only .000539.

We can modify the mean value extension (4.19) to obtain the *monotonicity test form*, $F_{MT}(X)$, as follows. Let S be the set of integers i (indices) such that

$\underline{D_i F(X)} < 0 < \overline{D_i F(X)}$; then we define

(4.26) $F_{\mathrm{MT}}(X) = [f(u), f(v)] + \sum\limits_{i \in S} D_i F(X)(X_i - m(X_i))$

where the pairs of real arguments (u_i, v_i) are defined by

(4.27) $(u_i, v_i) = \begin{cases} (\underline{X_i}, \bar{X_i}) & \text{if } \underline{D_i F(X)} \geqq 0, \\ (\bar{X_i}, \underline{X_i}) & \text{if } \underline{D_i F(X)} < 0 \text{ and } \overline{D_i F(X)} \leqq 0, \\ (m(X_i), m(X_i)) & \text{otherwise.} \end{cases}$

We have the following.

THEOREM 4.3. *For f in the class $FC_n(X_0)$ with $F(X)$ and $D_i F(X)$ ($i = 1, 2, \cdots, n$) expressible in the form (4.12) and extensions of its partial derivatives (see, e.g. (4.21)), we have $\bar{f}(X) \subseteq F_{\mathrm{MT}}(X)$ for all $X \subseteq X_0$.*

Proof. Without loss of generality, let $S = \{i : j \leqq i \leqq n\}$. Write $m_i = m(X_i)$. We have

(4.28) $f(x_1, x_2, \cdots, x_{j-1}, m_j, \cdots, m_n) \in [f(u), f(v)]$

for all $x_i \in X_i$, $i = 1, 2, \cdots, j-1$. Here, $u = (u_1, u_2, \cdots, u_n)$ and $v = (v_1, v_2, \cdots, v_n)$ where u_i and v_i are defined by (4.27). Now

(4.29) $f(x_1, \cdots, x_j, \cdots, x_n) - f(x_1, \cdots, m_j, \cdots, m_n) \in \sum\limits_{i \in S} D_i F(X)(X_i - m_i)$

for all $x_i \in X_i$, $i = 1, \cdots, n$. By adding (4.28) and (4.29) we conclude that $f(x) \in F_{\mathrm{MT}}(X)$ for all x in X, (N 7).

As we have seen in the previous example, the better the extensions of the partial derivatives, $D_i F(X)$, the better the results will be for the form $F_{\mathrm{MV}}(X)$. Compare (4.25) using (4.23) with (4.22) using (4.21).

Skelboe (N 8) has introduced an algorithm which can vastly reduce the number of evaluations required to compute refinements. Given an interval vector X and an interval extension $F(X)$ of a real valued function of n real variables, one can seek, first, a lower bound on the minimum value of f in X, by looking at a sequence of refinements. One evaluates f only on a finite subsequence of regions producing the smallest lower bounds. The process is repeated with $(-f)$ to obtain the upper bounds. While the interval extensions used for the evaluations can be any of the forms discussed here (or others) there is an advantage in using forms such as the centered form or the mean value form (or the monotonicity test modification of it) because of the more rapid convergence of the excess width to zero as the size of the regions grows small. Compared to straightforward use of the method of refinements (evaluating on *every* part of a subdivision) the algorithm of Skelboe is vastly superior, reducing the number of function evaluations in one two dimensional example from 10^{12} down to 85 to achieve an excess width of about 10^{-12}.

FINITE CONVERGENCE 49

Skelboe (N 8) proves that, in computing refinements, no subdivision is necessary of argument intervals for arguments which occur only once in the expression used for the interval extension. We present next a slight simplification of Skelboe's algorithm, using *cyclic bisection*.

Suppose we are given an interval extension $F(X)$ of a function f in the class $FC_n(X_0)$ and suppose we want to bound the range of values of f in X_0 using refinements of F. We can proceed as follows. We first find a lower bound and then, replacing $F(X)$ by $-F(X)$ we repeat the procedure to be described to find an upper bound. If the evaluations are carried out in rounded interval arithmetic, the procedure *will converge in a finite number of steps*, (see § 4.2). To find a lower bound, we create an *ordered list* of items of the form $(Y, F(Y))$, ordered so that $(Y, F(Y))$ comes before $(Z, F(Z))$ in the list only if $\underline{F(Y)} \leq \underline{F(Z)}$. The interval extension to be used could be the mean value form, the centered form, the monotonicity test form, or any other interval extension with inclusion monotonicity. The better the extension, the fewer the number of steps to convergence.

(4.30)   *The cyclic bisection for range of values*:
   (1) set $b_0 = \overline{F(X_0)}$;
   (2) set $X = X_0$;
   (3) to begin with, the *list* is empty;
   (4) set $i = 1$;
   (5) bisect X in coordinate direction $i: X = X^{(1)} \cup X^{(2)}$;
   (6) set $b = \min\{\underline{F(X^{(1)})}, \underline{F(X^{(2)})}\}$;
   (7) if $b \leq b_0$, then take b_0 as the lower bound; replace $F(X)$ by $-F(X)$ and repeat the procedure from step (1) to obtain the upper bound;
        otherwise proceed with step (8);
   (8) cycle i (if $i < n$, then replace i by $i+1$; if $i = n$, then replace i by $i = 1$);
   (9) set $b_0 = b$;
   (10) enter the items $(X^{(1)}, \underline{F(X^{(1)})})$ and $(X^{(2)}, \underline{F(X^{(2)})})$ in proper order in the list;
   (11) set X = the argument (first member of the pair) of the *first* item in the *list* (with lowest $\underline{F(X)}$) and remove the item $(X, F(X))$ from the list;
   (12) return to step (5).

Note that step (5) means we replace X by two interval vectors,
$$X^{(1)} = (X_1, \cdots, X_{i-1}, X_i^{(1)}, X_{i+1}, \cdots, X_n)$$
$$X^{(2)} = (X_1, \cdots, X_{i-1}, X_i^{(2)}, X_{i+1}, \cdots, X_n)$$
where $X_i^{(1)} = [\underline{X_i}, (\underline{X_i} + \overline{X_i})/2]$ and $X_i^{(2)} = [(\underline{X_i} + \overline{X_i})/2, \overline{X_i}]$.

For straightforward application of the method of refinements (4.4) with uniform subdivision into N subintervals in each of n coordinates the number of evaluations of $F(X)$ required is N^n. On the other hand, at least for most functions with only a finite number of isolated extrema in X_0, a procedure such as that of cyclic bisection, (4.30), will produce bounds on the range of values, of comparable excess width, in about $C_n (\log_2 N)$ evaluations of $F(X)$ for some constant C independent of n and N. If an extremal point x happens to fall on one or more interior faces during the bisection process (x in $X^{(1)} \cap X^{(2)}$), the number of evaluations might go up by a factor 2^m, for some $1 \leq m \leq n$. For $n = 10$, $N = 1000$ we have $N^n = 10^{30}$; whereas $n \log_2 N \approx 100$ and $2^n n \log_2 N \approx 10^5$.

For functions f which have only one (or perhaps a few) isolated simple maximum or minimum point(s) in X, we can make use of interval versions of Newton's method to further reduce the number of evaluations required to obtain sharp bounds on the range of values of f in X (N 9).

An efficient procedure for obtaining bounds on the range of values of a polynomial $P(x)$ with interval or real coefficients when x lies in an interval X has been given, complete with a triplex-Algol program, by Dussel and Schmitt, (N 10).

4.5. Integration. For continuous functions f, we have, from the mean value theorem,

$$(4.31) \qquad \int_a^b f(t)\, dt = \int_{[a,b]} f(t)\, dt = f(s)(b-a) \quad \text{for some } s \text{ in } [a, b].$$

It follows that,

$$(4.32) \qquad \int_X f(t)\, dt \in \bar{f}(X) w(X)$$

as long as f is continuous for t in X.

For f in $FC_1(X_0)$ and F an inclusion monotonic, Lipschitz, interval extension of f with $F(X)$ defined for all $X \subseteq X_0$, we have, since $\bar{f}(X) \subseteq F(X)$,

$$(4.33) \qquad \int_X f(t)\, dt \in F(X) w(X) \quad \text{for all } X \subseteq X_0.$$

Let N be a positive integer and subdivide $[a, b] \subseteq X_0$ into N subintervals $X_1, X_2, \cdots, X_N$ so that

$$a = \underline{X_1} < \overline{X_1} = \underline{X_2} < \overline{X_2} < \cdots < \overline{X_N} = b.$$

From the additive property of integration, we find that

$$(4.34) \qquad \int_{[a,b]} f(t)\, dt = \sum_{i=1}^N \int_{X_1} f(t)\, dt.$$

Applying (4.33) to (4.34), we have the following.

THEOREM 4.4. *There is a constant L, not depending on N or the method of subdivision, such that*

$$(4.35) \qquad \int_{[a,b]} f(t) \, dt \in \sum_{i=1}^{N} F(X_i) w(X_i)$$

and

$$w\left(\sum_{i=1}^{N} F(X_i) w(X_i) \right) \leq L \sum_{i=1}^{N} w(X_i)^2.$$

Proof. From § 4.1, for the Lipschitz extension F there is a constant L such that $w(F(X)) \leq L w(X)$ for all $X \subseteq [a, b] \subseteq X_0$. The result follows.

For a uniform subdivision of $[a, b]$, with $w(X_i) = (b-a)/N$ for $i = 1, 2, \cdots, N$, define

$$(4.36) \qquad S_N = S_N(F; [a, b]) = \sum_{i=1}^{N} F(X_i)(b-a)/N.$$

From Theorem 4.4, we have

$$(4.37) \qquad w(S_N) \leq L(b-a)^2/N.$$

As a result, we have the following.

THEOREM 4.5.

$$(4.38) \qquad \int_{[a,b]} f(t) \, dt = \bigcap_{N=1}^{\infty} S_N(F; [a, b]) = \lim_{N \to \infty} S_N(F; [a, b]).$$

It follows from Lemma 4.4 that the sequence of intervals defined by $Y_1 = S_1$, $Y_{k+1} = S_{k+1} \cap Y_k$, $k = 1, 2, \cdots$, is a nested sequence of intervals converging to the exact value of the integral.

If we evaluate $S_k^*(F; [a, b])$ in rounded interval arithmetic, then the sequence defined by $Y_1^* = S_1^*$, $Y_{k+1}^* = S_{k+1}^* \cap Y_k^*$, $k = 1, 2, \cdots$, *converges in a finite number of steps* to an interval containing the exact value of the integral.

We can use Theorem 4.5 to *define* the integral of an interval valued function of a real variable t.

A Lipschitz interval extension F (§ 4.1) has the property that $F(t)$ is real valued for real arguments t; however, for other useful, inclusion monotonic interval functions (for instance interval polynomial enclosures, see § 3.5) we may have interval number coefficients so that the width of $F(t)$ is not zero for real t.

Suppose that an interval valued function with values $F(X)$ is continuous and inclusion monotonic for $X \subseteq X_0$. If $[a, b] \subseteq X_0$, then the sums $S_N(F; [a, b])$ given in (4.36) are defined for F. For real arguments t (degenerate intervals

$[t, t]$) the values of $F(t)$ may be real numbers *or intervals*. We *define* the *interval integral*

(4.39) $$\int_{[a,b]} F(t)\, dt = \bigcap_{N=1}^{\infty} S_N(F;[a,b]).$$

It follows from the continuity of F that there are two continuous real valued functions $\underline{F}$, and $\bar{F}$ such that, for *real t*, we have $F(t) = [\underline{F}(t), \bar{F}(t)]$. It is not hard to show (N 11) that the integral defined by (4.39) is equivalent to

(4.40) $$\int_{[a,b]} F(t)\, dt = \left[\int_{[a,b]} \underline{F}(t)\, dt, \int_{[a,b]} \bar{F}(t)\, dt \right].$$

It is also easy to see that *interval integration preserves inclusion*:

(4.41)   if $F(t) \subseteq G(t)$ for all $t \in [a, b]$ then

$$\int_{[a,b]} F(t)\, dt \subseteq \int_{[a,b]} G(t)\, dt.$$

It follows, in particular, that if G is an interval polynomial enclosure of f, then

(4.42) $$\int_{[a,b]} f(t)\, dt \in \int_{[a,b]} G(t)\, dt.$$

More generally, if f is real valued and G is interval valued, continuous, and inclusion monotonic, with $f(t) \in G(t)$ for all t in $[a, b]$, then (4.42) holds.

We will now illustrate these results with some examples. Consider $f(t) = 1/t$ and put $F(X) = 1/X$. From (4.33) we have

(4.43) $$\int_{[1,2]} (1/t)\, dt \in (1/[1, 2]) = [\tfrac{1}{2}, 1].$$

From (4.36), we have

$$S_N = \sum_{i=1}^{N} (1/(1 + [i-1, i]/N))/N$$

$$= \sum_{i=1}^{N} [1/(N+i), 1/(N+i-1)].$$

Using interval arithmetic rounded to three decimal digits, we obtain

$$S_1^* = [.5, 1.0],$$
$$S_2^* = [.583, .834],$$
$$S_3^* = [.617, .783],$$
$$\cdots,$$
$$S_{10}^* = [.663, .722],$$
$$\cdots.$$

For the exact value of S_N, we have the actual widths $w(S_N) = 1/N - \frac{1}{2}N = 1/(2N)$ for this example, so the sequence $Y_{k+1}^* = S_{k+1}^* \cap Y_k^*$ with $Y_1^* = S_1^*$ will converge (using thee decimal digit rounded interval arithmetic) in no more than 500 steps to an interval of width no less than .001 containing the exact value of the integral. This is slow convergence for such an easy integral; we will introduce faster methods for bounding integrals in this section. First, however, an example is needed to illustrate the application of (4.42) and (4.39). Suppose we wish to find upper and lower bounds on the integral

$$(4.44) \qquad\qquad I = \int_0^1 e^{-t^2} \, dt.$$

For the function under the integral in (4.44) we have the interval polynomial enclosure P defined by

$$(4.45) \qquad\qquad e^{-t^2} \in P(t) = 1 - t^2 + \tfrac{1}{2}t^4 - [.0613, .1667]t^6$$

for all t in $[0, 1]$. The interval coefficient $[.0613, .1667]$ bounds the range of values of $\frac{1}{6} e^{-s^2}$ for all s in $[0, 1]$. (See also § 3.5.) From (4.42) we find, for I in (4.44),

$$(4.46) \qquad\qquad I \in \int_{[0,1]} P(t) \, dt.$$

It follows from (4.40) that (using (4.45))

$$I \in \int_{[0,1]} P(t) \, dt = [1 - \tfrac{1}{3} + \tfrac{1}{10} - .1667/7, \ 1 - \tfrac{1}{3} + \tfrac{1}{10} - .0613/7]$$
$$(4.47)$$
$$\subseteq [.742, .758].$$

By using narrower interval polynomial enclosures, we can, as we shall see, compute arbitrarily sharp bounds on the exact value of integrals such as (4.44).

It is important to note that, when carrying out formal integration of polynomials with interval coefficients, the end points of a coefficient $A_i = [\underline{A}_i, \bar{A}_i]$ will be switched upon multiplication by a negative real number. This results in the following two cases. Let $A_0, A_1, \cdots, A_q$ be intervals.

(4.48) *Case 1. If a and b have the same sign, then*

$$\int_{[a,b]} (A_0 + \cdots + A_q t^q) \, dt = A_0(b-a) + \cdots + A_q(b^{q+1} - a^{q+1})/(q+1);$$

     Case 2. If a and b have opposite signs, with $a < 0 < b$ then

$$\int_{[a,b]} (A_0 + \cdots + A_q t^q) \, dt = T_0 + T_1 + \cdots + T_q$$

where $T_i = A_i(b^{i+1} - a^{i+1})/(i+1)$, for i *even*; and

$$T_i = [(\underline{A}_i b^{i+1} - \bar{A}_i a^{i+1})/(i+1), (\bar{A}_i b^{i+1} - \underline{A}_i a^{i+1})/(i+1)],$$

for i *odd*. We can obtain the same result by adding the integrals over $[a, 0]$ and $[0, b]$.

Note that for "real" (degenerate interval) coefficients A_i, with $\underline{A}_i = \bar{A}_i$, both cases reduce to the usual formal integration of polynomials.

We can formally differentiate a given polynomial with interval coefficients as well. However, it is *not* true that the resulting polynomial contains the derivatives of all functions enclosed by the given polynomial, (N 2). If we have an interval polynomial enclosure of the derivative of a function, say $f'(t) \in P(t)$, then formal integration of $P(s)$ from a to t will produce an interval polynomial enclosure of $f(t) - f(a)$ whose formal derivative, namely $P(t)$ encloses $f'(t)$. Thus, the formal integral of an interval polynomial enclosure of the derivative of a function f is called a *derivative compatible enclosure of f*, (N 13). It is *not* a derivative compatible enclosure of *every* function it contains.

To illustrate the concept introduced in the previous paragraph, we reconsider the polynomial enclosure $P(t)$ (4.45). By formal integration from 0 to t, we obtain the interval polynomial

(4.49) $$Q(t) = t - \tfrac{1}{3}t^3 + \tfrac{1}{10}t^5 - ([.0613, .1667]/7)t^7.$$

The polynomial $Q(t)$ is a derivative compatible enclosure of the function

(4.50) $$q(t) = \int_0^t e^{-s^2}\, ds \quad \text{for } t \text{ in } [0, 1];$$

that is to say, we have $q(t) \in Q(t)$ and $q'(t) \in Q'(t)$ where $Q'(t) = P(t)$ is the formal derivative of $Q(t)$.

By using the techniques for recursive evaluation of derivatives given in § 3.4, we can efficiently compute bounds on the exact values of integrals. From (3.16) and (4.34), (4.48), we have, for any positive integers K and N, if a and b have the *same sign*,

THEOREM 4.6.

(4.51)
$$\int_{[a,b]} f(t)\, dt \in \sum_{i=1}^{N} \left\{ \left(\sum_{k=0}^{K-1} (1/(k+1))(f)_k(\underline{X}_i)w(X_i)^{k+1} \right) \right.$$
$$\left. + (1/(K+1))(f)_K(X_i)w(X_i)^{K+1} \right\}$$

for functions f with inclusion monotonic interval extensions of f and its first K derivatives. This includes the class of functions $FC_n(X_0)$ defined in § 4.4. From the Lipschitz property of the interval extensions, there is a constant L_K such that

(4.52) $$w((f)_K(X_i)) \leq L_K w(X_i) \quad \text{for all } X_i \subseteq [a, b] \subseteq X_0.$$

If we denote the right hand side of (4.51) by $I_{N,K}$, then we have

(4.53) $$w(I_{N,K}) \leq (L_K/(K+1)) \sum_{i=1}^{N} w(X_i)^{K+2}.$$

For a uniform subdivision of $[a, b]$ we have $w(X_i) = (b-a)/N$ for all $i = 1, \cdots, N$ and so (4.53) becomes

(4.54) $w(I_{N,K}) \leq C_K h^{K+1}$ where $C_K = L_K(b-a)/(K+1)$ and $h = (b-a)/N$.

Thus, for fixed K, the right hand side of (4.51), which we denote by $I_{N,K}$, contains the exact value of the integral and is an interval of width no more than some constant C_K times the $(K+1)$st power of the "step size" $h = (b-a)/N$.

For the first example considered in this section (4.43) it is not hard to show that we can take $L_K = K+1$ and so we can take $C_K = 1$ for all K. Thus, $I_{N,K}$ for the application of (4.51) to (4.43) is, for uniform subdivision, an interval of width no more than $(1/N)^{K+1}$ for any positive integers N and K. In this example we have, furthermore, $(f)_k(\underline{X}_i) = (-1)^k/\underline{X}_i^{k+1}$ and $(f)_K(\overline{X}_i) = (-1)^K/\overline{X}_i^{K+1}$, with $\underline{X}_i = 1 + (i-1)/N$, $w(X_i) = 1/N$, and $X_i = 1 + [i-1, i]/N$. To make the width of $I_{N,K}$ be .001 or less, we can take $N \geq 10^{3/(K+1)}$. The simpler method (4.36) with width bounded by (4.37) corresponds to putting $K = 0$ in (4.51) and deleting the sum over k. The following pairs of integers satisfy $N \geq 10^{3/(K+1)}$.

(4.55)

K	N
0	1000
1	32
2	10
3	6
4	4
5	4
6	3
7	3
8	3
9	2

For any $K > 9$ we will still need $N = 2$.

Since the exact value of the integral is contained in $I_{N,K}$ for *every* pair of integers $N \geq 1$, $K \geq 0$, we can intersect the intervals $I_{N,K}$ for *any* sequence of pairs of integers $\{(N, K)\}$ and obtain an interval which still contains the exact value of the integral. If we carry out all the computations in $I_{N,K}$ using rounded interval arithmetic, then *any* sequence of such intersections will *converge in some finite number of steps* to an interval containing the exact value of the integral. We could fix N and take an increasing sequence of values of K, or fix K and take an increasing sequence of values of N, for instance. Some work has been done on the question of optimal choices for the parameters N and K; however, there are still open problems in this area, (N 14).

For the second example considered in this section, the integral given by (4.44), we can also apply (4.51) to compute $I_{N,K}$ for any $N \geqq 1$, $K \geqq 0$, using the recursion formulas

$$(f)_0(X) = e^{-X^2},$$

(4.56) $$(f)_1(X) = -2X(f)_0(X),$$

$$(f)_k(X) = -(2/k)(X(f)_{k-1}(X) + (f)_{k-2}(X)) \quad \text{for } k \geqq 2.$$

For $N = 2$ and $K = 6$, using rounded interval arithmetic, we find $I_{N,K} = I_{2,6} = [.746, .748]$. Compare this with (4.47).

We can also apply (4.51) to more complicated functions in $FC_n(X_0)$, e.g. $x(t)$ in (3.23) using the recursion relations (3.25) for the derivatives, (N 15).

It is important to note that, during the computation of $I_{N,K}$ using (4.51) in rounded interval arithmetic, we must also compute the coefficients $(f)_k(X_i)$ in rounded interval arithmetic in order to ensure that the computed interval value of $I_{N,K}$ contains the value defined by (4.51).

Interval extensions of a variety of numerical integration (quadrature) methods have been programmed and discussed by J. H. Gray and L. B. Rall (N 15). These include Newton–Cotes formulas of both open and closed type, Gaussian quadrature, and Euler–Maclaurin integration. The programs find bounds, as needed, for remainder terms by recursive evaluation of Taylor coefficients in interval arithmetic using the methods of § 3.4. The entire process is automated, so that only the integrand need be programmed—in ordinary FORTRAN notation.

NOTES

1. See W. Strother [94] and also R. E. Moore [57].

2. R. E. Moore [57].

3. See K. Nickel and K. Ritter [71], K. Nickel [68], [69] and P. Wisskirchen [99] for further discussion of finite convergence and its applications.

4. See R. E. Moore [62], [59]; F. Bierbaum [70, pp. 160–168]; H. Ratschek [70, pp. 48–74]; K.-U. Jahn [37], [38], [39].

5. E. Hansen [25, 102–106]; see also H. Ratschek [80] and R. E. Moore [57].

6. G. Alefeld and J. Herzberger [1a, Satz 6].

7.   See R. E. Moore [60] for a discussion of uses of monotonicity tests for reducing the number of evaluations needed to obtain bounds on the range of values of a function.

8.   S. Skelboe [88].

9.   R. E. Moore [60]; see also R. Dussel [70, pp. 169–177].

10.   R. Dussel and B. Schmitt [14].

11.   R. E. Moore [57].

12.   H. Ratschek and G. Schroeder [81], have presented a thorough discussion of the concept of derivative for interval valued functions. See also G. Schroeder [87].

13.   "ableitungsverträgliche Einschliessung"; see F. Krückeberg [46], and also K. Kansy [41].

14.   R. E. Moore [76, Vol. I, pp. 61–130], [59] and also J. H. Gray and L. B. Rall [20].

15.   J. H. Gray and L. B. Rall [20], [21]; see also R. E. Moore [59], [76, Vol. I, pp. 61–130], [57], and K. Nickel [66].

Chapter 5

Computable Sufficient Conditions for Existence and Convergence

5.1. Linear algebraic systems. In this section we consider finite systems of linear algebraic equations

$$(5.1) \qquad\qquad Ax = b$$

where A is an $n \times n$ matrix and b is an n-dimensional vector. There are two cases to consider: (1) the coefficients of A and b are real numbers exactly representable by machine numbers, or (2) the coefficients of A and b are only known to lie in certain *intervals*, A_{ij} and B_i.

There are two types of methods for the numerical solution of such problems. So-called *direct* methods, such as Gaussian elimination (with or without various "pivoting" schemes) can produce exact results in case (1) in a finite number of arithmetic operations if A is nonsingular and if infinite precision arithmetic is used. *Indirect* (or iterative) methods for the solution of (5.1) in case (1) produce a sequence of approximate solutions which will converge to the unique solution if, for instance, A is of the form $A = I - M$ where I is the identity matrix and M has norm less than 1, for instance for the norm (maximum row sum norm)

$$(5.2) \qquad \|M\| = \max_i \sum_j |M_{ij}|, \quad \text{(a special case of (2.8))}.$$

An exact solution in case (2) is much more difficult; the set of all solutions to (5.1), when the coefficients of A and b can range over intervals, may be a very complicated set. It can, however, be *enclosed* in an interval vector (an n-dimensional rectangle) (N 1). As an immediate consequence of Corollary 3.1 in Chapter 3, we have the following computable *test for existence* of solutions to (5.1), with real or interval coefficients.

THEOREM 5.1. *If we can carry out all the steps of a direct method for solving (5.1) in rounded interval arithmetic (if no attempted division by an interval containing zero occurs nor any overflow or underflow), then the system (5.1) has*

59

*a unique solution for every real matrix in A and every real vector in b and the
solution is contained in the resulting interval vector X.*

In case (1), with exactly representable real coefficients, if the hypothesis of
Theorem 5.1 is satisfied, the resulting interval vector will have a width which
goes to zero as the number of digits carried (in the rounded interval arithmetic)
increases. We illustrate this phenomenon with an example. Consider the
following ill-conditioned system.

$$2.000 \, x_1 + 3.001 \, x_2 = 1.000$$

$$.6667 \, x_1 + 1.000 \, x_2 = .3333.$$

We will carry out a direct method for solving the system using rounded interval
arithmetic with n decimal digits, for $n = 4, 5, 6, 7, 8, 9$.

We can eliminate the x_1 term in the second equation to obtain

$$(1.000 - (.6667/2.000)(3.001))x_2 = .3333 - (.6667/2.000)(1.000).$$

If we carry out the arithmetic using rounded interval arithmetic with n decimal
digits, we obtain (for x_2):

for n = 4: $.6667/2.000 \in [.3333, .3334]$,

$\quad\quad\quad\quad (.6667/2.000)(3.001) \in [1.000, 1.001]$,

$\quad\quad\quad\quad .3333 - (.6667/2.000)(1.000) \in [-.0001, 0]$,

$\quad\quad\quad\quad (1.000 - (.6667/2.000)(3.001)) \in [-.0010, 0]$,

$\quad\quad\quad\quad x_2 \in [0, \infty)$ (no upper bound on x_2);

for n = 5: $.6667/2.000 \in [.33335, .33335]$,

$\quad\quad\quad\quad (.6667/2.000)(3.001) \in [1.0003, 1.0004]$,

$\quad\quad\quad\quad .3333 - (.6667/2.000)(1.000) \in [-.00005, -.00005]$,

$\quad\quad\quad\quad (1.000 - (.6667/2.000)(3.001)) \in [-.00040, -.00030]$,

$\quad\quad\quad\quad x_2 \in [.12500, .16667]$;

for n = 6: $x_2 \in [.128205, .131579]$;

for n = 7: $x_2 \in [.1302083, .1305484]$;

for n = 8: $x_2 \in [.13041210, .13044613]$;

for n = 9: $x_2 \in [.130429111, .130429112]$.

The sudden increase in accuracy in this simple example which occurs
between $n = 8$ and $n = 9$ is accounted for by the fact that, beyond $n = 8$, the
only roundoff error remaining is in the final division for x_2. In this example, the

width of the computed interval goes down by, at least, 10^{-1} for each increase of a unit in n beyond $n = 5$.

A phenomenon, somewhat puzzling at first, was observed during early experiments with interval versions of Gaussian elimination. In case (1), with real coefficients, the midpoint of the resulting interval solution in rounded interval arithmetic is usually very much closer to the exact solution than the width of the interval solution—by seven or eight decimal places for systems of order around twenty. This phenomenon has been studied and is now well understood, (N 2). For a fixed number of digits carried in the rounded interval arithmetic operations, the width of the resulting interval vector will increase as the order n increases for most matrices at about the same rate as the worst case analysis of von Neumann and Goldstine. During the Gaussian elimination process there are multiple occurrences of the coefficients in combination with subtractions and divisions which reverse endpoints, resulting in the observed excess width (in most cases) of the interval results. For special classes of matrices, interval Gaussian elimination can give good results. It has been shown recently (N 3) that interval Gaussian elimination always can be carried out, even without pivoting, when A is an "M matrix"; or when A is strongly, diagonally dominant, (N 4). (See also H. Beeck (N 3).)

Investigations of interval versions of indirect or *iterative* methods were begun by E. Hansen and subsequently pursued by a number of other researchers (N 5). We can multiply both sides of (5.1) by a matrix Y (for instance an approximate inverse of $m(A)$) and define $E = I - YA$. If $\|E\| < 1$ using (2.8), then the sequence

(5.3) $$X^{(k+1)} = \{Yb + EX^{(k)}\} \cap X^{(k)}, \quad k = 0, 1, 2, \cdots,$$

with

$$X_i^{(0)} = [-1, 1]\|Yb\|/(1 - \|E\|), \qquad i = 1, 2, \cdots, n,$$

is a nested sequence of interval vectors containing the unique solution to (5.1) for every real matrix in A and every real vector in b. In rounded interval arithmetic, the sequence (5.3) will converge in a finite number of steps to an interval vector containing the set of solutions to (5.1). Thus, we have:

THEOREM 5.2. *Using the norm* (2.8), *the system* (5.1) *has a unique solution x for every real matrix in A and every real vector in b (for interval matrix A and interval vector b) if $\|I - YA\| < 1$ for some matrix Y (real or interval). Furthermore, the solution vector x is contained in the interval vector $X^{(k)}$ defined by* (5.3) *for every $k = 0, 1, \cdots$. Using rounded interval arithmetic, the sequence $\{X^{(k)}\}$ converges in a finite number of steps to an interval vector containing the set of solutions to* (5.1).

We can illustrate (5.3) with a simple example. Consider (5.1) with

$$A = \begin{pmatrix} 3 & 1 \\ 3 & 2 \end{pmatrix}, \qquad b = \begin{pmatrix} 1 \\ 0 \end{pmatrix}.$$

If we choose

$$Y = \begin{pmatrix} .6 & -.3 \\ -1 & 1 \end{pmatrix}, \quad \text{an approximate inverse of } A,$$

we find, using three decimal digit rounded interval arithmetic,

$$E = \begin{pmatrix} .1 & 0 \\ 0 & 0 \end{pmatrix}, \quad Yb = \begin{pmatrix} .6 \\ -1 \end{pmatrix}, \quad X^{(0)} = \begin{pmatrix} [-1.12, & 1.12] \\ [-1.12, & 1.12] \end{pmatrix};$$

and we obtain

$$X^{(1)} = \begin{pmatrix} [.478, & .712] \\ -1 \end{pmatrix}, \quad X^{(2)} = \begin{pmatrix} [.647, & .672] \\ -1 \end{pmatrix}, \quad X^{(3)} = \begin{pmatrix} [.664, & .668] \\ -1 \end{pmatrix}$$

$$X^{(4)} = \begin{pmatrix} [.666, & .667] \\ -1 \end{pmatrix}$$

with $X^{(k)} = X^{(4)}$ for $k \geq 4$.

The iterative method (5.3) is also applicable in case A and b have interval components. Sharper bounds on the set of solutions may be obtained using methods of E. Hansen or G. Alefeld (N 1). As already mentioned (following (5.2)) the exact set of solutions, when A and b have interval components, may be a complicated set. Even for a 2×2 interval matrix A and a 2-dimensional interval vector b, the set of solutions to

$$A_r x = b_r$$

for $A_r \in A$, $b_r \in b$ can have a boundary which is a nonconvex, eight sided polygon. Such an example has been given by E. Hansen (N 1).

5.2. Finite dimensional nonlinear systems. In this section we consider finite systems of nonlinear equations

(5.4) $f_1(x_1, x_2, \cdots, x_n) = 0,$

$f_2(x_1, x_2, \cdots, x_n) = 0,$

$\cdots$

$f_n(x_1, x_2, \cdots, x_n) = 0$

which we may write, in vector notation,

(5.5) $f(x) = 0.$

As in the previous section, we can consider two cases: (1) the functions f_i are exactly representable real valued functions, or (2) the functions f_i have coefficients known only to lie in certain intervals. We will discuss case (1) first and then extend the results to include case (2). (See also E. Adams and W. F. Ames (N 16).)

Suppose that f in (5.5) is continuously differentiable in an open domain D; f and f' have inclusion monotonic interval extensions F and F', for instance the united extensions, defined on interval vectors $X \subseteq D$. We have the following *computational test for the existence of a solution* (N 6).

THEOREM 5.3. *Let Y be a nonsingular real matrix and let y be a real vector contained in the interval vector $X \subseteq D$. Define $K(X)$ by*

$$(5.6) \qquad K(X) = y - Yf(y) + \{I - YF'(X)\}(X - y).$$

If $K(X) \subseteq X$, then (5.5) *has a solution in X. It is also in $K(X)$.*

The proof of Theorem 5.3 is straightforward (N 6). If the interval vector $X = (X_1, X_2, \cdots, X_n)$ is an n cube, so that $w(X_i) = w(X)$ for $i = 1, 2, \cdots, n$, and if we choose $y = m(X)$, then $K(X)$ will lie in the interior of X if

$$(5.7) \qquad \|K(X) - m(X)\| < w(X)/2.$$

Thus, for an n cube X, (5.7) is sufficient for the existence of a solution to (5.5) in X. The same condition (5.7), which can be verified computationally by evaluating $K(X)$, is also sufficient to guarantee the convergence of a variety of iterative algorithms as indicated by the following.

THEOREM 5.4. *Let X be an n cube, $y = m(X)$, and Y a nonsingular real matrix. Suppose that* (5.7) *is satisfied. Put $X^{(0)} = X$, $Y^{(0)} = Y$ and consider an arbitrary real vector $x^{(0)}$ in $X^{(0)}$. Then the system* (5.4) *has a unique solution in X and the following algorithms converge to the solution:*

$$(5.8) \qquad x^{(k+1)} = x^{(k)} - Yf(x^{(k)}), \qquad k = 0, 1, 2, \cdots;$$

$$(5.9) \qquad x^{(k+1)} = x^{(k)} - Y^{(k)}f(x^{(k)}), \ k = 0, 1, 2, \cdots \text{ as long as}$$
$$\|I - Y^{(k)}F'(X)\| \leq r \qquad \qquad \text{for some } r < 1;$$

$$(5.10) \qquad X^{(k+1)} = X^{(k)} \cap K(X^{(k)}), \quad k = 0, \ 1, \ 2, \ \cdots, \quad \text{where } K(X^{(k)}) = y^{(k)} - Y^{(k)}f(y^{(k)}) + \{I - Y^{(k)}F'(X^{(k)})\}Z^{(k)}, \text{ and } y^{(k)} = m(X^{(k)}), Z^{(k)} = X^{(k)} - m(y^{(k)}) \text{ and where } Y^{(k)} \text{ is chosen as follows}$$

$$Y^{(k)} = \begin{cases} Y, \text{ an approximation to } [m(F'(X^{(k)}))]^{-1}, \\ \qquad \text{if } \|I - YF'(X^{(k)})\| \leq \|I - Y^{(k-1)}F'(X^{(k-1)})\|; \\ Y^{(k-1)} \quad \text{otherwise.} \end{cases}$$

Proof. The importance of Theorem 5.4 warrants inclusion of a proof even though it has appeared (and in more detail) elsewhere (N 7). For existence, we have the following. If $P(x) = x - Yf(x)$ maps X into itself, then $f(x) = 0$ has a solution x^* in X. This follows, for nonsingular Y, from the continuity of f and the fact that X is convex and compact, by the Schauder fixed point theorem. It is not hard to see that $P(x) \in K(X)$, (N 7). Thus $K(X) \subseteq X$ implies that P maps X into itself. Furthermore, the condition (5.7), which implies that $K(X) \subseteq X$, also implies that

$$\|I - YF'(X)\| < 1.$$

This can be shown as follows. Since X is an n cube, we have

$$X_i - m(X_i) = [-1, 1]w/2 \quad \text{for all } i = 1, 2, \cdots, n.$$

Thus

$$\{(I - YF'(X))(X - m(X))\}_i = \left\{ \sum_{j=1}^{n} |I - YF'(X)|_{ij} \right\}[-1, 1]w/2.$$

It follows that, for *some i*, we have

$$K(X)_i - m(X_i) = -[YF(m(X))]_i + \|I - YF'(X)\|[-1, 1]w/2.$$

From $|K(X)_i - m(X_i)| < w/2$ it follows that $\|I - YF'(X)\| < 1$. Since $P(x) \in K(X)$ for all x in X, and since $K(X) \subseteq X$, it follows by induction, from $x^{(0)} \in X$, that $x^{(k)} \in X$ for *all* k with $x^{(k)}$ defined by (5.8). Furthermore,

$$\|x^{(k+1)} - x^*\| \le \|I - YF'(X)\| \, \|x^{(k)} - x^*\|$$

since $f(x^{(k)}) - f(x^*) \in F'(X)(x^{(k)} - x^*)$. Therefore, from $\|I - YF'(X)\| < 1$, it follows that the sequence defined by (5.8) converges to the *unique* solution x^* (of (5.5)) in X from any $x^{(0)}$ in X. The convergence of the algorithms (5.9) and (5.10) is proved in a similar way (N 7).

We will illustrate the use of the algorithms (5.8), (5.9), (5.10) with an example. Consider the system of equations

(5.11)
$$f_1(x_1, x_2) = x_1^2 + x_2^2 - 1 = 0,$$
$$f_2(x_1, x_2) = x_1 - x_2^2 = 0.$$

For f' we have the Jacobian matrix

(5.12)
$$f'(x) = \begin{pmatrix} 2x_1 & 2x_2 \\ 1 & -2x_2 \end{pmatrix}.$$

For the interval extensions F and F' we can take the natural interval extensions of the corresponding real rational functions, simply evaluating (5.11) and (5.12) in interval arithmetic. For $X = ([.5, .8], [.6, .9])$ we have

$$m(F'(X)) = \begin{pmatrix} 1.3 & 1.5 \\ 1 & -1.5 \end{pmatrix}.$$

As an approximate inverse of this matrix, we will take

(5.13)
$$Y = \begin{pmatrix} .43 & .43 \\ .29 & -.37 \end{pmatrix}.$$

Putting $y = (.65, .75) = m(X)$, we find from (5.6) for the 2-cube X,

$$K(X) \subseteq ([.559, .68], [.74, .84]).$$

Since $\|K(X) - m(X)\| = .091 < w(X)/2 = .15$ the hypotheses for Theorem 5.4 are satisfied and so the conclusions follow. In particular, the iterative method (5.8) converges to a solution of (5.11) using Y given by (5.13) from any $x^{(0)}$ in $X^{(0)} = ([.5, .8], [.6, .9])$. The other two Newton type algorithms (5.9) and (5.10) converge as well. The interval algorithm (5.10) produces a nested sequence of interval vectors containing the solution. Using rounded interval arithmetic, the algorithm (5.10) converges in a finite number of steps to an interval vector containing a solution of (5.11).

The algorithm (5.8) with Y chosen as $f'(x^{(0)})^{-1}$ is the *simplified Newton method*. The algorithm (5.9) with $Y^{(k)} = f'(x^{(k)})^{-1}$ is the *Newton method* for the system (5.4). The satisfaction of (5.7) is sufficient to guarantee the convergence of the simplified Newton method from any $x^{(0)}$ in X. The *interval Newton method* (5.10) produces a nested sequence of interval vectors containing the unique solution in X if (5.7) is satisfied. It can be shown (N 8) that the *widths* of the containing interval vectors *converge quadratically* to zero if $F'(X)$ is a Lipschitz extension of $f'(x)$. The ordinary Newton method will converge to the solution in X from any $x^{(0)}$ in X if $\|I - f'(y)^{-1}F'(X)\| < 1$ for all y in X.

The special case $n = 1$ in (5.4) has been considered by a number of authors. In the case of polynomial functions, methods for bounding the complex zeros in circles in the complex plane have been discussed by I. Gargantini and P. Henrici (N 9). *Globally convergent* interval Newton methods for finding all the real zeros of a continuously differentiable real valued function in an interval have been discussed by K. Nickel and by E. Hansen (N 10). The method of Hansen, which uses extended interval arithmetic (including *unbounded* intervals) and some new interval forms for derivatives, bounds *multiple* as well as simple roots.

A globally convergent method for the general n dimensional system does not seem to be available. In a later chapter we will discuss some search procedures for finding safe starting regions for iterative solution of (5.4). The computable sufficient conditions for existence and convergence given in Theorems 5.3 and 5.4 are much easier to use than the well known Kantorovich conditions (N 11).

In case (2) mentioned at the beginning of this section, when the functions f_i in (5.4) have coefficients known only to lie in certain intervals, we can still use the interval algorithm (5.10). In this case, we require inclusion monotonic interval functions F and F' such that $f(x)$ is contained in $F(x)$ for every choice of real constants in the interval coefficients and such that $f'(x)$ is contained in $F'(X)$ for every choice of real constants from the interval coefficients and for every x in X. We replace $f(y^{(k)})$ in (5.10) by $F(y^{(k)})$ and the condition (5.7) now implies, if it is satisfied, that the system (5.4) has a solution in X (perhaps a whole set of them) and that the sequence of interval vectors generated by (5.10), starting with $X^{(0)} = X$, $Y^{(0)} = Y$, is a nested sequence and converges in a finite number of steps (using rounded interval arithmetic) to an interval vector containing all the solutions to (5.4) in X *for any choice* of the real coefficients from the given interval coefficients. Thus we can compute bounds on the set of possible solutions in X.

The following simple one dimensional examples will serve to illustrate dramatically the difference between the ordinary Newton method and interval Newton methods. For a single equation we can use a simpler algorithm than (5.10). From the mean value theorem for continuously differentiable functions, we have

(5.14) $f(x) = f(y) + f'(s)(x - y)$ for some s between x and y.

If $f(x) = 0$, then we have

(5.15) $x = y - (f'(s))^{-1} f(y).$

Let $F'(X)$ be an inclusion monotonic interval extension of $f'(x)$ and consider the algorithm (N 12)

(5.16) $X^{(k+1)} = X^{(k)} \cap N(X^{(k)}),$ $k = 0, 1, 2, \cdots$

where

$$N(X) = m(X) - f(m(X))/F'(X).$$

From (5.15) it follows that x is contained in $N(X)$ if $y = m(X)$ and if x is contained in X for then s in (5.14) is also contained in X. In this case, x is also contained in $X^{(k)}$ for all k if it is contained in $X^{(0)}$. We have the following.

THEOREM 5.5. *If an interval $X^{(0)}$ contains a zero x of $f(x)$, then so does $X^{(k)}$ for all $k = 0, 1, 2, \cdots$, defined by (5.16); furthermore, the intervals $X^{(k)}$ form a nested sequence converging to x if $0 \notin F'(X^{(0)})$.*

Proof. If $0 \notin F'(X^{(0)})$, then $0 \notin F'(X^{(k)})$ for all k and the midpoint $m(X^{(k)})$ is not contained in $X^{(k+1)}$. Therefore we have $w(X^{(k+1)}) < \frac{1}{2} w(X^{(k)})$. The convergence of the sequence follows.

Actually, we can even allow 0 to be contained in $F'(X^{(0)})$ if we use extended interval arithmetic as the following example shows. Consider $f(x) = -2.001 + 3x - x^3$. We have $f'(x) = 3(1 - x^2)$. We can use the interval extension $F'(X) = 3(1 - X^2)$; see (3.10). If we put $X^{(0)} = [-3, 3]$, we obtain $F'(X^{(0)}) = [-24, 3]$ and, in extended interval arithmetic, we obtain

(5.17) $F'(X^{(0)})^{-1} = [-\infty, -\frac{1}{24}] \cup [\frac{1}{3}, \infty].$

Thus the interval $X^{(0)}$ is mapped here into the union of two disjoint, unbounded intervals. From this point we can follow the production of one (or more) sequences of intervals using the algorithm (5.16) beginning with either of the intervals in the union (5.17). In this example, if we choose the second interval in the union (5.17), the sequence of intervals generated by (5.16) will terminate in a finite number of steps with the empty interval since there is no zero of $f(x)$ beyond $\frac{1}{3}$. Conversely, since a zero of $f(x)$ is in $N(X)$ whenever it is in X, we can *tell* from an empty intersection $X \cap N(X)$ that there is no zero in X. If we choose the first interval in (5.17), then we find, using (5.16), a nested sequence of intervals

(5.18)

$$X^{(0)} = [-3, 3],$$

$$X^{(1)} = [-3, -.083375],$$

$$X^{(2)} = [-3, -1.66526],$$

$$X^{(3)} = [-2.17875, -1.63830],$$

$$X^{(4)} = [-2.17875, -2.06189],$$

$$X^{(5)} = [-2.0162, -2.0003],$$

$$X^{(6)} = [-2.00024, -2.00006],$$

$$X^{(7)} = [-2.0001112, -2.0001110].$$

During the computation, we found that $N(X^{(4)}) \subseteq X^{(4)}$, with $0 \notin F'(X^{(4)})$. It follows that $f(x)$ has a zero in $X^{(k)}$ for all k in this sequence.

THEOREM 5.6. *If* $0 \notin F'(X)$ *and* $N(X) \subseteq X$ *with* $N(X) = m(X) - f(m(X))/F'(X)$, *then* $f(x)$ *has a zero in* X.

Proof. (See E. Hansen (N 13) for a proof.)

In contrast to the behavior of the sequence found above, (5.18), the ordinary Newton method, $x^{(k+1)} = x^{(k)} - f(x^{(k)})/f'(x^{(k)})$ generates very erratic sequences for this example unless $x^{(0)}$ is less than -1. For instance, with $x^{(0)} = 0$, the ordinary Newton method generates the sequence

$$x^{(1)} = .667,$$

$$x^{(2)} = .84518716,$$

$$x^{(3)} = .92592529,$$

$$x^{(4)} = .965774,$$

$$x^{(5)} = .98794069,$$

$$x^{(6)} = 1.0078932,$$

$$x^{(7)} = .98291958,$$

$$x^{(8)} = 1.0013261,$$

$$x^{(9)} = .87506664,$$

$$x^{(10)} = .94034361,$$

$$\cdots$$

It is a commonly observed phenomenon that the ordinary Newton method can get hung up, oscillating near a place where a local extremum occurs if $f(x)$ has a very small nonzero value there. The interval Newton method (5.16) will reject such a region as not containing a zero by producing an empty intersection in (5.16).

An even more drastic difference between the interval and ordinary Newton methods is illustrated by the next example. Consider $f(x) = x^{1/3} = (\text{sgn } x)|x|^{1/3}$. We have $f'(x) = x^{-2/3}/3$. We can take $F'(X) = X^{-2/3}/3$ where

(5.19)
$$X^{-2/3} = \begin{cases} [\bar{X}^{-2/3}, \underline{X}^{-2/3}], & 0 < \underline{X} \leq \bar{X}, \\ [|X|^{-2/3}, \infty], & \underline{X} < 0 < \bar{X}, \\ [|\underline{X}|^{-2/3}, |\bar{X}|^{-2/3}], & \underline{X} \leq \bar{X} < 0. \end{cases}$$

Thus we have $X^{-2/3} = \{x^{-2/3} : x \in X\}$. We can deal with the unbounded interval using extended interval arithmetic. Consider first the interval Newton method (5.16) with *any* $X^{(0)}$ such that $\underline{X}^{(0)} < 0 < \bar{X}^{(0)}$. We find that $N(X) = m(X) - m(X)^{1/3}/(X^{-2/3}/3)$. We have, from (5.19), (5.16),

$$X^{(1)} = X^{(0)} \cap \{m(X^{(0)}) - 3m(X^{(0)})^{1/3}[0, |X^{(0)}|^{2/3}]\}.$$

We can distinguish three cases:
   (1) if $m(X^{(0)}) > 0$, then $X^{(1)} \subseteq [X^{(0)}, m(X^{(0)})]$;
   (2) if $m(X^{(0)}) < 0$, then $X^{(1)} \subseteq [m(X^{(0)}), \bar{X}^{(0)}]$;
   (3) if $m(X^{(0)}) = 0$, then $X^{(1)} = 0$.
In the first two cases, the width of $X^{(1)}$ is no more than half that of $X^{(0)}$. In any case, if $X^{(0)}$ contains 0, then *the sequence* (5.16) *will contain and converge to the solution* $x = 0$. The width of $X^{(k)}$ will be no more than $2^{-k}w(X^{(0)})$. This result is in strong contrast with the behavior of the ordinary Newton method for this example. For the ordinary Newton method we have

$$\begin{aligned} x^{(k+1)} &= x^{(k)} - f(x^{(k)})/f'(x^{(k)}) \\ &= x^{(k)} - 3x^{(k)} \\ &= -2x^{(k)}. \end{aligned}$$

Thus, the ordinary Newton method, for this example, *diverges from any* $x^{(0)}$, except the solution itself.

It is possible to extend some of the theorems and algorithms of this section to operator equations in function spaces. We will leave this as an *open problem*.

In this section we have considered interval versions of Newton's method for the iterative solution of (5.4). The interval Newton operators $N(X)$ and $K(X)$ given by (5.16) and (5.6) are *not*, in general, inclusion monotonic interval functions. The approach of the next section is based on inclusion monotonic interval functions. We will obtain a useful interval version of the contraction mapping theorem. The results of the next section can be applied to (5.4) if we first rewrite the system of equations as a fixed point problem.

5.3. Operator equations. Recall from § 3.5 that an *interval enclosure* of a real valued function f is an *inclusion monotonic* interval valued function F such that $f(t) \in F(t)$. Interval polynomial enclosures are particularly useful. Recall, also, from § 4.5, that we can formally integrate interval polynomials (4.48) and that interval integration preserves inclusion (4.41). In this section we will consider operator equations of the form

$$(5.20) \qquad\qquad y(t) = p(y)(t)$$

where the operator p may include derivatives and integrals of the function $y(t)$. We will consider some interval methods for establishing the existence of solutions to (5.20) and computational tests for existence of solutions and for convergence of iterative methods for approximate solution of (5.20). For clarity of notation, we will restrict our attention in this section to equations of the form (5.20) in which we seek a real valued function $y(t)$ of a single real variable t. The methods can be extended easily to the case of a vector valued function of a real variable as in systems of ordinary differential equations; and some of the methods can be extended to cover vector valued functions of vector valued t (e.g. systems of partial differential equations).

If $X(t)$ and $Y(t)$ are interval (or interval vector) valued functions with a common domain, we will write

$$(5.21) \quad X \subseteq Y \quad \text{if} \quad X(t) \subseteq Y(t) \quad \text{for all } t \text{ (in the common domain)}.$$

Similarly, if $x(t)$ is a real (or real vector) valued function, we write

$$(5.22) \qquad\qquad x \in X \quad \text{if} \quad x(t) \in X(t) \quad \text{for all } t.$$

Suppose the operator p in (5.20) is defined for some class M_r of real valued functions y with common domain $a \le t \le b$ and that $p: Mr \to M_r$. Let the *interval operator* $P: M \to M$ be defined on a class M of interval enclosures of elements of M_r with $M_r \subseteq M$. We call P an *interval majorant* of p if

$$(5.23) \qquad\qquad p(y) \in P(Y) \quad \text{for } y \in Y.$$

An interval operator P is *inclusion monotonic* if

$$(5.24) \qquad\qquad X \subseteq Y \quad \text{implies} \quad P(X) \subseteq P(Y).$$

We can usually write down such an operator P immediately, given p. (See Examples 1 and 2 in this section.)

For example, if H and F are inclusion monotonic (see (3.4)), then the interval operators of the form

$$(5.25) \qquad P(Y)(t) = H(t, Y(t)) + \int_a^{b(t)} F(t, s, Y(s)) \, ds$$

with $b(t) = b$ or $b(t) = t$ are inclusion monotonic because of (4.41).

The following theorem (N 14) provides a basis for useful computational tests for existence of solutions to (5.20) and for the convergence of iterative algorithms for solving operator equations.

THEOREM 5.7. *If P is an inclusion monotonic interval majorant of p, and if* $P(Y^{(0)}) \subseteq Y^{(0)}$, *then the sequence defined by*

$$(5.26) \qquad Y^{(k+1)} = P(Y^{(k)}), \qquad k = 0, 1, 2, \cdots,$$

has the following properties:
   (1) $Y^{(k+1)} \subseteq Y^{(k)}$, $k = 0, 1, 2, \cdots$;
   (2) *for every* $a \leqq t \leqq b$, *the limit*

$$(5.27) \qquad Y(t) = \bigcap_{k=0}^{\infty} Y^{(k)}(t)$$

   exists as an interval $Y(t) \subseteq Y^{(k)}(t)$, $k = 0, 1, 2, \cdots$;
   (3) *any solution of* (5.20) *which is in* $Y^{(0)}$ *is also in* $Y^{(k)}$ *for all k and in Y as well; that is, if* $y(t) \in Y^{(0)}(t)$ *for all* $a \leqq t \leqq b$, *then* $y(t) \in Y^{(k)}(t)$ *for all k and all* $a \leqq t \leqq b$;
   (4) *if there is a real number c, such that* $0 \leqq c \leqq 1$, *for which* $X \subseteq Y^{(0)}$, $(X \in M)$, *implies*

$$(5.28) \qquad \sup_{t} w(P(X)(t)) \leqq c \sup_{t} w(X(t)), \qquad a \leqq t \leqq b,$$

   then (5.20) *has the unique solution* $Y(t)$ *in* $Y^{(0)}$ *given by* (5.27).

Proof. Property (1) follows by induction from $Y^{(1)} = P(Y^{(0)}) \subseteq Y^{(0)}$, using the inclusion monotonicity of P. For any fixed t, the sequence $Y^{(k)}(t)$ of nested intervals converges to an interval $Y(t)$ which is expressible as the intersection in property (2). If y is a solution of (5.20), then $y \in Y^{(0)}$ implies $p(y) \in P(Y^{(0)})$ since P is an interval majorant of p; but $y = p(y)$, so $y \in Y^{(1)} = P(Y^{(0)})$. By induction we obtain property (3). From (5.28), it follows that the limit $Y(t)$ in (2) is real valued for every t and that $Y(t) = P(Y(t))$. From (5.23) it follows that $Y(t)$ is a solution to (5.20). Uniqueness in $Y^{(0)}$ follows from the *contraction property* (5.28). This completes the proof.

Note that, even without (5.28), we will have convergence of the sequence (5.26) to some *interval valued* function $Y(t)$.

Some examples will illustrate the application of Theorem 5.7.

Example 1. The initial problem

$$(5.29) \qquad dy/dt = t^2 + y(t)^2, \qquad y(0) = 0,$$

can be written in the form (5.20) with

$$(5.30) \qquad p(y)(t) = \int_0^t (s^2 + y(s)^2) \, ds.$$

We define in a natural way the interval operator P by

(5.31) $P(X)(t) = \int_0^t (s^2 + X(s)^2)\, ds$ (see § 4.5).

Let $Y^{(0)}(t) = [0, w]$ for $0 \leq t \leq b$; then we have

$$P(Y^{(0)})(t) = \int_0^t (s^2 + [0, w^2])\, ds$$
$$= t^3/3 + [0, w^2]t.$$

We will have $P(Y^{(0)}) \subseteq Y^{(0)}$ if $b^3/3 + w^2 b \leq w$. This will be the case if, for instance $w = .5$ and $b = .9$; thus, since P is (by construction) an inclusion monotonic majorant of p (see (3.5) and (4.41)), we can satisfy the hypotheses of Theorem 5.7 for this choice of w and b.

The operators p and P are defined for continuous functions y and X. From (5.31) we find that, for $0 \leq t \leq b$ and $X \subseteq T^{(0)} = [0, w]$,

(5.32) $w(P(X)(t)) = \int_0^t (\bar{X}(s)^2 - \underline{X}(s)^2)\, ds \leq 2bw \sup_t w(X(t));$

therefore (5.28) is satisfied with $c = 2bw$ if $2bw \leq 1$. It turns out here that the b and w we found to satisfy $P(Y^{(0)}) \subseteq Y^{(0)}$, namely $w = .5$ and $b = .9$ also satisfy (5.28). In other examples, we might have to reduce b to satisfy (5.28) after we have found w and b to satisfy $P(Y^{(0)}) \subseteq Y^{(0)}$. It follows, from all this, that the initial value problem (5.29) has a unique solution (expressible as the limit of the convergent sequence (5.26)) at least for $0 \leq t \leq .9$. We can apply the procedure again with $t = .9$ and $Y(.9)$ as a new initial point to continue the solution beyond $t = .9$. In Chapter 8 we will go into more detail about the construction of bounds on solutions of operator equations.

Example 2. Consider the following two point boundary value problem arising in chemical reactor theory (N 15).

(5.33)
$$y'' + (1/t)y' + B \exp(-1/|y|) = 0,$$
$$y'(0) = 0, \qquad y(1) = T, \quad \text{for real numbers } T, B > 0.$$

We can rewrite (5.33) as $(ty')' = -Bt \exp(-1/|y|)$. Integrating this twice and using the boundary values, we get an integral equation of the form (5.20) with

(5.34) $p(y)(t) = T + B \int_t^1 (1/s) \int_0^s u \exp(-1/|y(u)|)\, du\, ds.$

If we take $Y^{(0)} = [T, \infty]$, then we find that, for all $0 \leq t \leq 1$,

(5.35) $P(Y^{(0)})(t) = T + B[\exp(-1/T), 1](1 - t^2)/4 \subseteq Y^{(0)}$

for the interval operator

$$(5.36) \qquad P(Y)(t) = T + B \int_t^1 (1/s) \int_0^s u \exp(-1/Y(u)) \, du \, ds.$$

It follows from property (3) of Theorem 5.7 that any solution of (5.33) is contained in $P(Y^{(0)})$. It has been shown by S. V. Parter (N 15) that (5.33) may have no solution, one solution, or more than one solution, depending on the values of T and B. We can apply property (4) of Theorem 5.7 to derive computable sufficient conditions for the existence of a unique solution to (5.33). From (4.19) it follows that

$$w(\exp(-1/Y(t))) \leqq |\exp(-1/P(Y^{(0)}))/P(Y^{(0)})^2| w(Y)(t) \quad \text{for } Y \subseteq P(Y^{(0)}).$$

We find that $\sup_t w(P(Y)(t)) \leqq c \sup_t w(Y(t)), 0 \leqq t \leqq 1$, for $Y \subseteq P(Y^{(0)})$ and for $c = B \exp(-1/(T + B/4))/(4T^2)$. Thus, the problem (5.33) will have a unique solution for B and T such that $(T + B/4) \log_e (B/(4T^2)) < 1$.

The following result may also be of some interest.

THEOREM 5.8 *Under the hypotheses of Theorem 5.7 including (5.28), the ordinary Picard iteration method*

$$(5.37) \qquad y^{(k+1)}(t) = p(y^{(k)})(t), \qquad k = 0, 1, 2, \cdots,$$

converges to the unique solution of (5.20) in $Y^{(0)}$ from any $y^{(0)}$ in $Y^{(0)}$.

Proof. From $y^{(0)} \in Y^{(0)}$ it follows that $y^{(k)} \in Y^{(k)}$ for all k with $Y^{(k)}$ defined by (5.26). In fact, $y^{(k-1)} \in Y^{(k-1)}$ implies $y^{(k)} = p(y^{(k-1)}) \in P(Y^{(k-1)}) = Y^{(k)}$. Since $\{Y^{(k)}\}$ converges to the unique solution of (5.20) in $Y^{(0)}$, so does $\{y^{(k)}\}$.

NOTES

1. A method of E. Hansen described in [25, pp. 35–45] and in [26], makes use of information concerning monotonicity. The method is a forerunner of our "monotonicity test form" (4.26); see also G. Alefeld [2] H. Beeck [70].

2. See J. von Neumann and H. Goldstine [97]. P. L. Richman and A. J. Goldstein [82]; and also W. Chuba and W. Miller [10], and W. Miller [54], [55], [70, pp. 246–249], H. Ratschek [78].

3. See G. Alefeld [3], M. Hegben [31], [32], H. Beeck [70, pp. 150–159].

4. See G. Alefeld and J. Herzberger [1a, § 16].

5. See E. Hansen [23], E. Hansen and R. Smith [30], F. Krückeberg [45], R. E. Moore [57, § 5.2], R. Krawczyk [43], G. Alefeld and J. Herzberger [1a], P. Wongwises [70, pp. 316–325], P. Wisskirchen [99]. In the paper of Wongwises it is shown both by theoretical argument and by extensive numerical experimentation that, for matrices with real coefficients, the excess width of the converged interval vector using iterative methods like (5.3) increases only very slowly or not at all with the order of the matrix.

6. For a proof of Theorem 5.3, see R. E. Moore [61]. Interval versions of Newton's method have been studied previously by R. E. Moore [56], [57]; K. Nickel [25, pp. 19–23], [67]; E. Hansen [25, p. 23], [24], R. Krawczyk [43]; K. Madsen [50], G. Alefeld and J. Herzberger [1a]. Computer programs based on interval Newton methods with automatic generation of the derivatives in the Jacobian matrix have been described by L. B. Rall (et. al.) [47], [19], [77].

7. See R. E. Moore [63], [61].
8. See R. E. Moore and S. T. Jones [64]; see also K. Madsen [50].
9. See I. Gargantini and P. Henrici [17], I. Gargantini [16].
10. See E. Hansen [27], [28], also K. Nickel [67], [25, pp. 10–24].
11. See L. V. Kanterovich, *Functional analysis and applied mathematics*, Uspehi Math. Nauk, 3 (1948), pp. 89–185; see also J. M. Ortega and W. C. Rheinboldt [74]; L. B. Rall [77a].
12. R. E. Moore [56], [57].
13. E. Hansen [25, p. 23].
14. R. E. Moore [70, pp. 31–47], [62].
15. R. E. Moore [62], S. V. Parter [75].
16. E. Adams and W. F. Ames [1].

Chapter 6

Safe Starting Regions
for Iterative Methods

If an iterative method for solving a *linear* system converges at all, it generally converges from *any* initial approximation. For example, if $\|E\| < 1$, then $x^{(k+1)} = b + Ex^{(k)}$ produces a sequence $\{x^{(k)}\}$ converging to the solution of $x = b + Ex$ from *any* $x^{(0)}$. The situation for *nonlinear* systems is quite different. In general, an iterative method for solving a nonlinear system of equations will converge to a solution only from initial approximations which are fairly close to a solution. While a linear system with real coefficients can have, at most, one *isolated* solution, a nonlinear system can have any number of isolated solutions, depending on the particular nonlinear system. By an "isolated solution" we mean a solution which has a neighborhood containing no other solutions. Clearly, even in one dimension, nonlinear equations (for instance polynomial equations) can have several isolated solutions. This can cause difficulties for iterative methods at points in between. There are many sources of difficulties for iterative methods for nonlinear systems (see the example following Theorem 5.6 for instance). A practical problem is that of *finding* a safe starting point from which an iterative method will converge to a solution of a nonlinear system. For some systems it may be easy; for others it is extremely difficult. In this chapter we will consider some applications of interval analysis to the design of search procedures for finding safe starting points for iterative methods.

We can make use of the computational test (5.7) for existence of a solution to a nonlinear system (5.4) in an n-dimensional cube X. If the test is satisfied, then the iterative methods (5.8), (5.9), and (5.10) converge to a unique solution in X. One search procedure for finding such a *safe starting region* X is based on the following.

THEOREM 6.1. *Let $E = (E_1, E_2, \cdots, E_n)$ be an n-cube symmetric about the origin. Thus, $E_i = [-1, 1]r$ for some $r > 0$, $i = 1, 2, \cdots, n$. Suppose that $\{x^{(k)}\}$ is a sequence of vectors which converges to a solution x of (5.4), computed any way whatsoever. Suppose further that the sequence of real matrices $\{Y^{(k)}\}$ in (5.9) or (5.10) converges to a real matrix Y^* such that $\|I - Y^*F'(x + E)\| < 1$. Then the* *computational test* (5.7) *for existence and convergence will eventually (for some k) be satisfied for* $X = x^{(k)} + E$, $Y = Y^{(k)}$.

Proof. Recall that $K(X)$, defined by (5.6), using X and Y as given in the theorem, will have the form, choosing $y = x^{(k)}$,

$$(6.1) \qquad K(x^{(k)} + E) = x^{(k)} - Y^{(k)}f(x^{(k)}) + \{I - Y^{(k)}F'(x^{(k)} + E)\}E.$$

For the symmetric n-cube E, we have $m(E) = 0$ and $w(E) = 2r$. We wish to show that, for some k, we will have

$$(6.2) \qquad \|K(x^{(k)} + E) - x^{(k)}\| < r.$$

Since, for any $\varepsilon_1 > 0$, there is a $\bar{k}$ such that $k \geq \bar{k}$ implies

$$\|\{I - Y^{(k)}F'(x^{(k)} + E)\}E\| < (\|I - Y^*F'(x + E)\| + \varepsilon_1)r,$$

it follows that (because of the convergence of the sequences $\{x^{(k)}\}$ and $\{Y^{(k)}\}$ and the continuity of f and F') for some k we will have (6.2).

The *search procedure* implied by Theorem 6.1 is the following:

(6.3)   Along with the computation of *any* sequence $x^{(k)}$ designed to produce approximate solutions of (5.4) we test the condition (6.2); if, for some k, the condition (6.2) is satisfied, then a solution to (5.4) exists in $x^{(k)} = E$ and the iterative methods (5.8), (5.9), and (5.10) converge for $X^{(0)} = x^{(k)} + E$. We can *try* any $r > 0$.

COROLLARY TO THEOREM 6.1. *If* $Y^* = f'(x)^{-1}$ *and* $\|f'(x) - F'(x + E)\| \leq Lr$, *then the search will be successful for* $r < 1/(L\|Y^*\|)$.

We will describe, next, another search procedure, based on successive bisections of n-dimensional rectangles. The search procedure to be given can be used with any iterative method for solving (5.4) if a computational test is provided for existence and convergence. Let $K(X)$ be defined by (5.6) and let X be an n-dimensional interval vector; consider the computational test

$$(6.4) \qquad K(X) \subseteq X \quad \text{and} \quad \|I - YF'(X)\| < 1.$$

It can be shown that (6.4) is sufficient for the existence of a unique solution of (5.4) in X and for the convergence of the iterative algorithms (5.8), (5.9), and (5.10), (N 1). If X is an n-cube, then we may use the test (5.7) instead.

We begin with some *exclusion tests*. If, for any i, 0 is not in the range of values of $f_i(x)$ for x in X, then X cannot contain a solution of (5.4). Since $F_i(X)$ contains the range of values of $f_i(x)$ for x in X, we have the following exclusion test.

(6.5)   If $0 \notin F_i(X)$ for some $1 \leq i \leq n$, then (5.4) has no solution in X.

Note that, for each i, $F_i(X)$ is an interval; thus we will have $0 \notin F_i(X)$ if $F_i(X)$ has a positive left endpoint or a negative right endpoint.

It is not hard to show (N 2) that if X contains a solution to (5.4) then so does $K(X)$. Thus, whenever we compute $K(X)$, we have available another exclusion test.

(6.6)   If $K(X) \cap X$ is empty, then there is no solution of (5.4) in X.

The intersection $K(X) \cap X$ will be *empty* if, for any i, we have

(6.7) $$\underline{K(X)_i} > \bar{X}_i \quad \text{or} \quad \overline{K(X)_i} < \underline{X}_i.$$

When using the search procedure to be described with a particular iterative method for solving (5.4), there may be other exclusion tests available. In any case, independent of the particular iterative method, we will always have the exclusion test (6.5).

We describe now a *recursive interval bisection search procedure*, given by R. E. Moore and S. T. Jones (N 3).

Let B be an *arbitrary* interval vector in the domain of F, an inclusion monotonic interval extension of f in (5.5). Suppose we have an iterative algorithm for solving (5.5) which has a set C of computable sufficient conditions for the existence of a solution of (5.5) in an interval vector $X \subseteq B$ and for the convergence of the algorithm from X to a solution of (5.5). We will say that $C(X)$ is *true* if all the conditions are met for X; otherwise, $C(X)$ is *false*. The algorithm concerned may operate with real vectors, in which case "convergence of the algorithm from X" can be interpreted as meaning convergence of the algorithm from any $x^{(0)}$ in X. If a real vector must be selected from X, we can take the midpoint $m(X)$. If E is a set of exclusion criteria for the iterative method concerned, we will say that $E(X)$ is *true* if we can exclude X, by one of the tests, as not containing a solution of (5.5); otherwise $E(X)$ is false. For any $X \subseteq B$, we will have one of the following three possibilities:

(1) $C(X)$ is true;
(2) $E(X)$ is true;
(3) $C(X)$ and $E(X)$ are both false.

It could happen that $C(X)$ or $E(X)$ cannot be evaluated for some $X \subseteq B$. In that case, the relevant statement is considered false. Thus, in particular, $C(X)$ is true only if the iterative algorithm concerned is defined on X. If neither $C(X)$ nor $E(X)$ can be evaluated, then they are both "false" and we have the third possibility above. For example, if $C(X)$ requires the inverse of a Jacobian matrix or an approximation to one, but the inverse does not exist or the approximation cannot be found (for *whatever* reason), then $C(X)$ is false.

Determination of which of the conditions (1), (2), (3) holds for X will be called *analysis of X*. The goal of the search procedure is to find an X such that $C(X)$ is true.

A *push-down stack T* is created during the search procedure. This is a "last-in, first-out" list. By *push X*, we will mean the operation of adding X to the top of the stack. By *pop X*, we will mean the operation of removing the top region in the stack (the last one added) and designating it as X. The *bisection search procedure* is as follows:

(6.8)   Step 1. Set the stack T to empty; set $X = B$; continue.

Step 2. Analyze X; if $C(X)$ is true, terminate search (X is a safe starting region); if $E(X)$ is true, exclude X and go to Step 5; otherwise continue.

Step 3. Bisect X; select half to analyze; push other half on stack T; continue.

Step 4. Set X = half selected to analyze; go to Step 2.

Step 5. If stack T is empty, then terminate search (there is no solution in B); otherwise, pop X from stack T; go to Step 2.

We discuss next some bisection rules. In Step 3, in which coordinate direction should we bisect X? Having decided this, which of the resulting two halves should we choose to analyze? Before we discuss these questions, we remark that, in practise, it might happen (because of the finite representation of machine numbers) that we cannot further bisect a small region X. In this case, we can print out such a region for possible further analysis using higher machine precision and continue the search as if the region had been excluded.

Bisection rules (I). Make a cyclic choice of coordinate direction in successive bisections; select the left half of the bisected interval.

For this first set of bisection rules, we bisect the interval vector $X = (X_1, X_2, \cdots, X_n)$ in coordinate direction x_i to obtain the two halves $X^{(1)}$, $X^{(2)}$ ($X = X^{(1)} \cup X^{(2)}$) with

$$
\begin{aligned}
X_j^{(1)} = X_j^{(2)} = X_j, \qquad j \neq i, \\
(6.9) \qquad X_i^{(1)} = [\underline{X}_i, m(X_i)], \qquad X_i^{(2)} = [m(X_i), \bar{X}_i].
\end{aligned}
$$

We always choose $X^{(1)}$ to analyze next; $X^{(2)}$ is pushed on the stack. On successive executions of Step 3, the index i is increased by one until $i = n$ is reached; after which, i is reset to 1. Thus we repeatedly cycle i through the indices $1, 2, \cdots, n$. For these bisection rules we will have $w(X) = 2^{-p} w(B)$ after np bisections, unless the procedure terminates before np bisections. If the procedure terminates, then one of two results will have been obtained: (1) there is no solution in B, (2) we have found a safe starting region. The procedure will terminate in a finite number of steps if, for all sufficiently small regions, either $C(X)$ is true or $E(X)$ is true. If $F(X)$ is a Lipschitz extension of $f(x)$, then $E(X)$ will be true for all sufficiently small regions X which do not contain a solution of (5.5). It is more difficult to ensure that $C(X)$ will be true for all sufficiently small regions X which do contain a solution. We will leave the discussion of the simple, but arbitrary, bisection rules (I) at this point and consider other sets of bisection rules designed specifically for the iterative algorithms (5.8), (5.9) and (5.10).

Bisection rules (II). Bisection in a coordinate direction x_i in which the component X_i is of maximum width; select a half toward which a Newton step points from $m(X)$.

With these bisection rules we will still have $w(X) = 2^{-p} w(B)$ after np bisections. However, the particular order in which the coordinate directions are chosen in successive bisections is not yet completely specified. Suppose the direction x_i is chosen. The choice of a half "toward which a Newton step points" means we will take $[\underline{X}_i, m(X_i)]$ if $-[Yf(m(X))]_i < 0$, or $[m(X_i), \bar{X}_i]$ if $-[Yf(m(X))]_i > 0$, or either if $-[Yf(m(X))]_i = 0$ for the ith component of the half of X selected to be analyzed next. The other components $(X_j, j \neq i)$ are left unchanged. If we cannot compute $Yf(m(X))$, we select the left half of X in direction i. We push the remaining half of X on the stack. *If*, for a particular sequence of bisections using rules (II), the Newton steps from the midpoints always point to a half containing a solution, then the sequence of midpoints of the selected halves in the sequence of bisections will converge to a solution. As a result of Theorem 6.1 and its corollary, it seems likely that the search (6.8), using (6.4) for $C(X)$ and (6.5), (6.6) for $E(X)$, will terminate in a finite number of steps in this case. The determination of precise conditions under which the search procedure (6.8) will terminate in a finite number of steps using bisection rules of type (II) will be left as an open problem.

> *Bisection rules* (III). Since $K(X)$ contains a solution of (5.5) whenever X does, we can intersect $K(X)$ with X in Step 3 of the search procedure (6.8) when using it with the iterative algorithms (5.8), (5.9), or (5.10). More specifically, with (6.4) for $C(X)$ and (6.5), (6.6)–(6.7) for $E(X)$ (in Step 2 of (6.8)), we can use bisection rules (II) in Step 3 but select $X' \cap K(X)$ to analyze in Step 4 and push $X'' \cap K(X)$ on the stack, where X' is the half of X selected by bisection rules (II) and where X'' is the other half of X. Note that we will only arrive at Step 3 with a region X if $E(X)$ is false. Now $E(X)$ is true if and only if we have either $0 \notin F(X)$ or $X \cap K(X)$ empty. Thus we will arrive at Step 3 with a region X only if $X \cap K(X)$ is nonempty. It follows that $X' \cap K(X)$ will be nonempty. It could happen that $X'' \cap K(X)$ is empty. If this is the case, then we do not push anything onto the stack. After np bisections we will obtain a region X of width $w(X) \leq 2^{-p} w(B)$.

Example 1. To illustrate the search procedure (6.8) just outlined with bisection rules (III) we reconsider the system of equations (5.11) and search the initial region $B = ([0, 1], [0, 1])$ for a safe starting region for the iterative algorithms (5.8)–(5.10). We have

$$F_1(X_1, X_2) = X_1^2 + X_2^2 - 1,$$

$$F_2(X_1, X_2) = X_1 - X_2^2,$$

(6.10)

$$F'(X) = \begin{pmatrix} 2X_1 & 2X_2 \\ 1 & -2X_2 \end{pmatrix}.$$

We analyze B and find that both $C(B)$ and $E(B)$ are false, so we bisect B. According to the bisection rules (II) we bisect B in coordinate direction x_1 (this is the first direction in which the maximum component width of B occurs). A Newton step from $m(B)$ points to the right, so we choose the right half of the bisected region. Thus we obtain $X' = ([.5, 1], [0, 1])$. When we intersect this with $K(B)$ we obtain $X = ([.5, 1], [.125, 1])$; we stack $X'' \cap K(B)$. We next analyze X. We find that $C(X)$ and $E(X)$ are both false, so we bisect X. After going through the same procedure we obtain, as the new region X to analyze, $X = ([.5, .98125], [.5625, 1])$. The previous bisection was in the x_2 direction, since that was the widest component of the region being analyzed. For this new X we find that $K(X) \subseteq X$ and $\|I - YF'(X)\| = .6190455 < 1$. The search procedure terminates with this X as a safe starting region for the iterative methods of interest. In fact the *ordinary Newton method* (5.9) converges rapidly from $m(K(X)) = (.62409084, .79004314)$, in our safe starting region. In *two iterations* of the ordinary Newton method from this starting point we obtain an approximate solution of the system (5.11) *accurate to at least eight decimal places* for x_1, and x_2, namely $x_1 = .61803399$, $x_2 = .78615138$. In this example, we have found a safe starting region in *two bisections* of B.

Another set of bisection rules we would like to consider is motivated by the following:

(6.11)   (1) Bisect X in a direction x_i for which $-[Yf(m(X))]_i \neq 0$ and $(x - m(X))_i$ has the same sign as $-[Yf(m(X))]_i$ *for some solution* x of $f(x) = 0$;

(2) choose the half toward which $-[Yf(m(X))]_i$ points; that is $X = X' \cup X''$ and we select X' as our new candidate (saving X'' for possible analysis later) where:

(a) $X' = ([\underline{X}_1, \bar{X}_1], \cdots, [m(X_i), \bar{X}_i], \cdots, [\underline{X}_n, \bar{X}_n])$ if $-[Yf(m(X))]_i > 0$, or

(b) $X' = ([\underline{X}_1, \bar{X}_1], \cdots, [\underline{X}_i, m(X_i)], \cdots, [\underline{X}_n, \bar{X}_n])$ if $-[Yf(m(X))]_i < 0$.

Note that if $-[Yf(m(X))]_i = 0$ for all $i = 1, 2, \cdots, n$ then we have found a solution $f(m(X)) = 0$, since Y is a nonsingular matrix.

The condition (6.11) will guarantee that the half X' chosen still contains a solution x *assumed* here to be in X. The bisection rules (1), (2) say, in effect, to pick a coordinate direction and bisected half toward which the ordinary Newton method moves from the midpoint of X.

Of course, we do not yet know that there is a solution in X; nor can we tell whether (6.11) is satisfied even if there is a solution in X. Therefore, we modify the ideal rules (1), (2) as follows:

Bisection rules (IV). (1)' bisect X in the first coordinate direction x_i for which $|[Yf(m(X))]_i| \geq |[Yf(m(X))]_j|$ for all $j = 1, 2, \cdots, n$;

(6.12)

(2)' use the same second rule as before, (2), where i is chosen by (1)'.

The bisection rules (1)', (2)' can be carried out computationally. Under certain conditions, we can show they will produce the same choice of X' as the ideal rules (1), (2).

THEOREM 6.2. *If X contains a solution x of $f(x) = 0$, then the bisection rules (1)', (2)' will select a half of X which still contains x if, for instance, a Newton step from $m(X)$, namely $m(X) - Yf(m(X))$, is close enough to x so that*

$$(6.13) \qquad \|x - (m(X) - Yf(m(X)))\| < \|Yf(m(X))\|.$$

Proof. If (6.13) is satisfied, then for the choice of direction satisfying (6.12) we have

$$\|Yf(m(X))\| = |[Yf(m(X))]_i|$$

and, for this i in particular, we have

$$|(x - m(X))_i - (-[Yf(m(X))]_i)| < |[Yf(m(X))]_i|.$$

Therefore, $(x - m(X))_i$ has the same sign as $-[Yf(m(X))]_i$.

THEOREM 6.3. *If X contains a solution x of $f(x) = 0$ and Y is a nonsingular real matrix, then the bisection rules (1)', (2)' will select a half of X which still contains x if $\|I - YF'(X)\|$ is sufficiently small.*

Proof. Since

$$x - (m(X) - Yf(m(X))) \in \{I - YF'(X)\}(x - m(X)),$$

we have, since $z \in Z$ implies $\|z\| < \|Z\|$,

$$\|x - (m(X) - Yf(m(X)))\| \leq \|I - YF'(X)\| \|x - m(X)\|.$$

Now (6.13) will be satisfied if

$$(6.14) \qquad \|I - YF'(X)\| \|x - m(X)\| < \|Yf(m(X))\|.$$

If $f(m(X)) = 0$, then $m(X)$ is a solution; otherwise $f(m(X)) \neq 0$ and, from the nonsingularity of Y, we have $\|Yf(m(X))\| > 0$. Thus (6.13) is satisfied if

$$(6.15) \qquad \|I - YF'(X)\| < \|Yf(m(X))\|/\|x - m(X)\|$$

and the theorem is proved.

If $m(X)$ is near x and if Y is close to $f'(x)^{-1}$, then the right hand side of (6.15) will be close to 1.

Using the bisection rules (IV) on Example 1, (6.10), a safe starting region was found in four bisections of B. We could modify the rules (IV) to make use of the intersection $X' \cap K(X)$ just as we did in the modification (rules (III)) of rules (II). Since $K(X)$ contains a solution x whenever X does, the Theorems 6.2 and 6.3 will still apply to such a modification. However we will now illustrate the application of the rules (IV) without the modification.

Example 2. Consider the nonlinear system of equations

(6.16)
$$f_1(x_2, x_2) := \sin x_1 + \cos x_2 - 1 = 0,$$
$$f_2(x_1, x_2) := 3 - 2 \cos x_1 - 2 \cos x_2 = 0.$$

Again let us take $B = ([0, 1], [0, 1])$ and seek a solution in B—or rather, a safe starting region within B for an iterative algorithm for computing a solution. We can put

(6.17)
$$F'(X) = \begin{pmatrix} \cos X_1 & -\sin X_2 \\ 2 \sin X_1 & 2 \sin X_2 \end{pmatrix}.$$

For $X = B$, we have

$$m(X) = \begin{pmatrix} .5 \\ .5 \end{pmatrix},$$

$$F'(X) \subseteq \begin{pmatrix} [.54, 1] & [-.8415, 0] \\ [0, 1.683] & [0, 1.683] \end{pmatrix}.$$

$$m(F'(X)) \approx \begin{pmatrix} .77 & -.42075 \\ .8415 & .8415 \end{pmatrix},$$

$$f(m(X)) \approx \begin{pmatrix} .3570081 \\ -.5103303 \end{pmatrix}.$$

We find an approximate inverse of $m(F'(X))$, namely

$$Y = \begin{pmatrix} .8398 & .4199 \\ -.8398 & .76845 \end{pmatrix} \approx [m(F'(X))]^{-1}$$

and then

$$-Yf(m(X)) \approx \begin{pmatrix} -.08552771 \\ .69197872 \end{pmatrix},$$

$$K(X) \subseteq \begin{pmatrix} [-.2122, 1.0411] \\ [.81852, 1.5655] \end{pmatrix} \not\subseteq X.$$

The bisection rules (IV) choose the new region

$$X' = \begin{pmatrix} [0, 1] \\ [.5, 1] \end{pmatrix}.$$

Continuing with the search procedure using the bisection rules (IV), we obtain the numerical results shown in the following table.

X	$K(X)$	$-Yf(m(X))$
$\begin{pmatrix} [0, 1] \\ [0, 1] \end{pmatrix}$	$\begin{pmatrix} [-.2, 1.04] \\ [.81, 1.56] \end{pmatrix}$	$\begin{pmatrix} -.08 \\ .69 \end{pmatrix}$
$\begin{pmatrix} [0, 1] \\ [.5, 1] \end{pmatrix}$	$\begin{pmatrix} [.065, .764] \\ [.756, 1.183] \end{pmatrix}$	$\begin{pmatrix} -.08 \\ .22 \end{pmatrix}$
$\begin{pmatrix} [0, 1] \\ [.75, 1] \end{pmatrix}$	$\begin{pmatrix} [.124, .704] \\ [.701, 1.192] \end{pmatrix}$	$\begin{pmatrix} -.08 \\ .07 \end{pmatrix}$
$\begin{pmatrix} [0, .5] \\ [.75, 1] \end{pmatrix}$	$\begin{pmatrix} [.357, .519] \\ [.880, 1.041] \end{pmatrix}$	$\begin{pmatrix} .188 \\ .085 \end{pmatrix}$
$\begin{pmatrix} [.25, .5] \\ [.75, 1] \end{pmatrix}$	$\begin{pmatrix} [.39, .457] \\ [.916, .974] \end{pmatrix}$	$\begin{pmatrix} .05 \\ .07 \end{pmatrix}$

Thus, after four bisections, we have found a safe starting region. For the final X in the table, we do have $K(X)$ *contained in the interior of X*, therefore (5.7) is satisfied and Theorem 5.4 can be applied. Thus $X = ([.25, .5], [.75, 1])$ is a safe starting region for the algorithms (5.8)–(5.10).

Example 3. Consider the two-point boundary value problem

(6.18) $$\frac{d^2 x}{dt^2} = e^{-x} \quad \text{with } x(0) = x(1) = 0.$$

We obtain the following nonlinear system of equations as a finite difference approximation to the boundary value problem. Let $x = (x_1, \cdots, x_n)$, where x_k approximates $x(kh)$. We have

(6.19) $f_k(x) = -2x_k - h^2 e^{-x_k} + x_{k+1} + x_{k-1} = 0, \qquad k = 1, 2, \cdots, n,$

with $x_0 = x_{n+1} = 0$ and $h = 1/(n+1)$.

We have here, for $k, j = 1, \cdots, n$,

(6.20) $$F'(X)_{kj} = \begin{cases} -2 + h^2 e^{-X_k}, & j = k, \\ 1, & |j - k| = 1, \\ 0, & \text{otherwise.} \end{cases}$$

Since a solution to (6.18), if it exists, is clearly negative except at the endpoints, we can try $B = ([-1, 0], \cdots, [-1, 0])$. We put $X = B$ to begin with, as usual, and compute $K(X)$. In this example, we can compute $b = -Yf(m(X))$ easily by solving the linear system $m(F'(X))b = -f(m(X))$ directly, making use of the tri-diagonality of the matrix $m(F'(X))$. In fact, we can express $K(X)$ as $K(X) = m(X) + b + C$ where the interval vector C can also be solved for directly from the linear system

$$m(F'(X))C = (m(F'(X)) - F'(X))(X - m(X))$$

using the multiplicative factors saved during the forward, upper tri-angularization process in the direct solution for b. We need only perform a back substitution process in interval arithmetic to find the components of the interval vector C. We have carried this out for the case $n = 9$, $h = .1$. The resulting components of $K(X)$ are all contained in the interior of the interval $[-1, 0]$; therefore $K(X)$ is contained in the interior of the n-cube ($n = 9$), $X = ([-1, 0], \cdots, [-1, 0])$. Thus we find that the B we began with is already a safe starting region for the solution of the nonlinear system (6.19) using the iterative algorithms (5.8)–(5.10). If we choose to use the ordinary Newton method (5.9) we do not need an explicit matrix representation of Y, but can solve for successive displacements $x^{(k+1)} - x^{(k)}$ by direct solution of linear systems with the same coefficient matrix as used in finding b and C. Thus each simplified Newton iteration would require only re-evaluation of the right hand side $(-f(x^{(k)}))$ and the back-substitution process. In this example, *no bisections* were required.

While the bisection search procedure can be carried out even if the Jacobian matrix of the system (5.4) is singular at various places in the region B being searched, the number of bisections required to find a safe starting region will increase with the smallness of the size of the safe starting region sought. The presence of singularities of the Jacobian matrix may force the size of a safe starting region to be quite small for *any* iterative method. The following system arises in the derivation of optimal quadrature formulas for multi-dimensional integration.

$$x_1 + x_3 + x_5 + 2x_7 = 1,$$

$$x_1 x_2 + x_3 x_4 + 2x_5 x_6 + 2x_7 x_8 + 2x_7 x_9 = \tfrac{2}{3},$$

$$x_1 x_2^2 + x_3 x_4^2 + 2x_5 x_6^2 + 2x_7 x_8^2 + 2x_7 x_9^2 = \tfrac{2}{5},$$

$$x_1 x_2^3 + x_3 x_4^3 + 2x_5 x_6^3 + 2x_7 x_8^3 + 2x_7 x_9^3 = \tfrac{2}{7},$$

(6.21) $\qquad$ $$x_1 x_2^4 + x_3 x_4^4 + 2x_5 x_6^4 + 2x_7 x_8^4 + 2x_7 x_9^4 = \tfrac{2}{9},$$

$$x_5 x_6^2 + 2x_7 x_8 x_9 = \tfrac{1}{9},$$

$$x_5 x_6^4 + 2x_7 x_8^2 x_9^2 = \tfrac{1}{25},$$

$$x_5 x_6^3 + x_7 x_8 x_9^2 + x_7 x_8^2 x_9 = \tfrac{1}{15},$$

$$x_5 x_6^4 + x_7 x_8 x_9^3 + x_7 x_8^3 x_9 = \tfrac{1}{21}.$$

It is readily seen that the system (6.21), which was obtained from P. Rabinowitz (private communication), has a Jacobian matrix which is singular on the 8-dimensional surfaces: (1) $x_2 = x_4$, (2) $x_8 = x_9$, (3) $x_1 = 0$, (4) $x_3 = 0$, (5) $x_5 = 0$, (6) $x_7 = 0$, and perhaps others. A solution is sought in the 9-cube $B = ([0, 1], [0, 1], \cdots, [0, 1])$. Actually, a solution to the problem is already known; however, the system is typical of other, higher dimensional problems of this type which still await numerical solution. The search procedures described in this chapter seem to require large number of bisections for such systems. For a slightly different B, a safe starting region was found (S. T. Jones, personal communication) in 168 bisections with 73 backtracks. In fact, there does not seem to be *any* method, at present, which has an easy time with such problems.

A comparison (N 3) of the bisection search procedure (6.8) with a continuation method shows better results for the bisection procedure.

The study of methods for finding safe starting regions or safe starting points for iterative solution of nonlinear systems of equations is still in its early stages. Much remains to be done.

NOTES

1. See, for example R. E. Moore [61], [63].
2. See R. Krawczyk [43], R. E. Moore [61].
3. R. E. Moore and S. T. Jones [64].

Chapter 7

Applications to Mathematical Programming (Optimization)

7.1. Linear optimization. A number of authors have considered applications of interval computation to problems of linear optimization (N 1). In particular, B. Machost and, later, R. Krawczyk have applied interval analytic methods to the simplex method for solving linear programming problems. From an approximate solution of a linear optimization problem, Krawczyk (N 1) obtains an interval vector which contains the exact solution, taking into account errors in initial data and rounding errors. In this section we will describe the method given by Krawczyk. (See referenced work (N 1) for proofs.) We refer to § 2.1 for notation.

Let A be an $m \times n$ matrix, $m < n$, with interval coefficients. Let B and P be, respectively, m and n dimensional interval vectors. We wish to find an interval vector Z which contains the *set* of solutions to the linear programming problems: for each $b \in B$, $p \in P$, $A_r \in A$:

(7.1)   maximize the *objective function* $Q(x) = (p, x)$, subject to the constraints

$$A_r x = b \quad (\text{or } A_r x \leqq b \text{ without "slack" variables}),$$

$$0 \leqq x.$$

Here (p, x) denotes the inner product of the real vectors p and x. Also b is a real vector and A_r is a real matrix.

Suppose $\bar{z}$ is an *approximate* solution of (7.1) for some particular $\bar{p} \in P$, $\bar{b} \in B$, $\bar{A}_r \in A$, obtained using the simplex method, ($\bar{z}$ may contain rounding errors). Let S be the index set of all the *basis* variables of the approximate solution $\bar{z}$. Computable sufficient conditions for the set of all solutions of (7.1) to have the same basis as $\bar{z}$ will be given later. Assume, for the moment, this is the case.

Denote by x' the m dimensional vector consisting of the basis components of an n dimensional vector x. Thus $x' = (x_{i_1}, \cdots x_{i_m})$ where $S = \{i_1, \cdots, i_m\}$. Similarly, denote by x'' the $n - m$ dimensional vector consisting of all the nonbasis components of x. Let A'_r denote the $m \times m$ matrix consisting of the basis columns of A_r and A''_r the $(n-m) \times m$ matrix consisting of the nonbasis columns of A_r, etc.

Suppose that z' is the solution of $A'_r z' = b'$ for some $A'_r \in A'$, $b' \in B'$. We assume that all $A'_r \in A'$ are nonsingular. Let Z' be the set of such solutions z' for all $A'_r \in A'$, $b' \in B'$. Thus, $\bar{z}'$ is an *approximate* solution of $A'_r z' = b'$. Let Y be an approximation to the inverse of the matrix $\bar{A}'_r$ which was used in the computation of $\bar{z}$. Krawczyk (N 1) proves the following, *assuming that* $R = \|I - YA'\| < 1$.

THEOREM 7.1. *The set of solutions to the linear programming problem* (7.1) *for all* $b \in B$, $p \in P$, $A_r \in A$ *is contained in the interval vector Z computed as follows*:

$$Z'_i = \bar{z}_i + q[-1, 1] \quad (basis\ components\ of\ Z),$$

(7.2) $\qquad\qquad Z'' = 0 \qquad\qquad (nonbasis\ components\ of\ Z)$

$$where\ q = (\|Y\| \|A'\bar{z}' - B\|)/(1 - R).$$

If $w(A)$ and $w(B)$ are small, the bound Z can be sharpened. We can find a narrower interval vector containing the set of solutions to (7.1) as follows. Call $Z^{(0)} = Z'$ (as found in Theorem 7.1). For $k = 0, 1, 2, \cdots$ compute

(7.3) $\qquad\qquad Z^{(k+1)} = Z^{(k)} \cap \{YB + (I - YA')Z^{(k)}\}.$

The iterations (7.3) yield a nested sequence of interval vectors containing the basis components of the set of solutions to (7.1). In rounded interval arithmetic the sequence (7.3) will converge in a finite number of steps. We again set the nonbasis components of Z equal to zero.

To determine whether the set of all solutions has the same basis as $\bar{z}$, we proceed as follows. We can easily test $Z' \geqq 0$. Next, we denote the transposes of the matrices A' and A'' by A'^T and A''^T, respectively. Let P' and P'' be the basis components and the nonbasis components, respectively, of the vector of objective function coefficients. By the methods of § 5.1, we can find an interval vector V containing the set of solutions of $A'^T_r v = p'$ for all $A'_r \in A'$ and all $p' \in P'$. If $A''^T V - P'' \geqq 0$, then the set of solutions has the same basis as $\bar{z}$; and we can then apply Theorem 7.1 and the algorithm (7.3).

For the maximum value of $Q(x)$, we have $Q(z) \in (P, Z)$.

7.2. Convex programming.

In this section we consider the following problem. Let f be a continuously differentiable real valued function of n real variables $x = (x_1, x_2, \cdots, x_n)$ in a domain including an n dimensional rectangle representable as an interval vector $B = (B_1, B_2, \cdots, B_n)$, $B_i = [\underline{B}_i, \bar{B}_i]$, $i = 1, \cdots, n$. We assume that f is strictly convex in B. We seek the point $\bar{x}$ in B such that

(7.4) $\qquad\qquad\qquad f(\bar{x}) = \min_{x \in B} f(x).$

The existence of a unique such point is well known and easy to prove. R. Dussel (N 2) has given an interval analytic method for finding the "smallest possible" interval vector X containing $\bar{x}$ and the "smallest possible" interval F containing $f(\bar{x})$. We will describe the method of Dussel.

We remark first that a sufficient condition for the strict convexity of f in B is

(7.5) $\partial^2 f(x)/\partial x_i \partial x_j > 0$ for all x in B and all $i, j = 1, \cdots, n$.

We denote the gradient of f at x by $f'(x)$. Thus $f'_i(x) = (\partial f/\partial x_i)(x)$. It is assumed that f is continuously differentiable in B. We assume also that f and f'_i, $i = 1, \cdots, n$, have inclusion monotonic, Lipschitz interval extensions F and F'_i with $F(X)$ and $F'_i(X)$ defined for all $X \subseteq B$. If $F'_i(X) > 0$ it follows that $f'_i(x) > 0$ for all x in X; similarly, $F'_i(X) < 0$ implies $f'_i(x) < 0$ for all x in X. We define the function

(7.6) $\operatorname{sign}(Z) = \begin{cases} 1 & \text{if } 0 < Z \ (0 < \underline{Z}), \\ 0 & \text{if } 0 \in Z \ (\underline{Z} \leq 0 \leq \bar{Z}), \\ -1 & \text{if } 0 > Z \ (0 > \bar{Z}) \end{cases}$

for an interval $Z = (\underline{Z}, \bar{Z})$. For degenerate intervals, this reduces to the ordinary sign function for real numbers.

We define the $(n-1)$-dimensional rectangles

$$H(i, b) = \{x : x \in B, x_i = b\}$$

for $i = 1, 2, \cdots, n$ and $\underline{B}_i < b < \bar{B}_i$.

We can represent $H(i, b)$ by the interval vector $(B_1, \cdots, b, \cdots, B_n)$ where $B_i = b$. The function f has a unique minimum point in each such $H(i, b)$. Dussel (N 2) proves the following.

THEOREM 7.2. *Suppose that* $f(y) = \min f(x)$ *for all x in $H(i, b)$ and $\bar{z}$ is defined by* (7.4); *then*

(7.7) $\operatorname{sign}(f'_i(y)) = \operatorname{sign}(b - \bar{x}_i)$.

If it should happen that $\operatorname{sign}(f'_i(y)) = 0$, then we would have $y = \bar{x}$; otherwise we have

(7.8) $\begin{aligned} \bar{x}_i < b \quad &\text{if } \operatorname{sign}(f'_i(y)) = 1, \\ \bar{x}_i < b \quad &\text{if } \operatorname{sign}(f'_i(y)) = -1. \end{aligned}$

The method of Dussel consists of a cyclic sequence of bisections of B (cyclic choice of coordinate directions), choosing the half at each bisection which still contains $\bar{x}$. We can determine which half this is by using (7.8). In order to compute $\operatorname{sign}(f'_i(y))$, we note that $\operatorname{sign}(F'_i(X)) = \pm 1$ implies $\operatorname{sign}(f'_i(y)) = \pm 1$ whenever y is in X. That is, if $\operatorname{sign}(F'_i(X)) \neq 0$, then $\operatorname{sign}(F'_i(X)) = \operatorname{sign}(f'_i(y))$ if X contains y. The problem of finding y to minimize $f(x)$ for x in $H(i, b)$ is again a problem of the type (7.4), but of dimension $n-1$. We do not have to find y accurately. We only need to know a region X containing y such that sign $(F_i'(X)) \neq 0$. We seek such a region by again applying cyclic bisection to the region $H(i, b)$. Dussel (N 2) gives a recursive procedure, written in Triplex-Algol 60, for carrying out the method.

The procedure can be carried out in rounded interval arithmetic with the bisections proceeding until, because of the finite representation of machine numbers, a machine computed midpoint of an interval is no longer in the interior of a computed interval vector containing $\bar{x}$. At this point the procedure is terminated and we have found the smallest possible interval vector containing $\bar{x}$ which is representable in the chosen set of machine numbers.

The procedure was tried out by Dussel (N 2) on numerous examples using the EL X8 at Karlsruhe University. One example, of dimension 4, was the following. For $f(x) = x_1^4 + \cosh x_2 + x_2^3 + x_4^4 + x_1^2 x_3^4 + x_2^2 x_4^2$ with $B = ([-2.1, 2.9], [-2.2, 2.8], [-2.3, 2.7], [-2.4, 2.6])$, an interval vector containing the exact solution of the minimization problem (7.4) was found of the form

$$\bar{x} \in X = ([-.108(10^{-11}), .497(10^{-13})], \cdots, [-.226(10^{-11}), .228(10^{-11})])$$

with $w(X) = .852(10^{-11})$. The minimum value of $f(x)$ was bounded by $F(X) = 1 + [-1.1(10^{-11}), 1.9(10^{-11})]$. The computation required 310 bisections, of which 171 were of the 4 dimensional region B and the rest were bisections of lower dimensional regions required to find signs of gradient components.

Note that, for an n-dimensional problem of type (7.4), after np cyclic bisections of B we will obtain a region X of width $w(X) = 2^{-p} w(B)$. Thus 164 cyclic bisections of B in the example just mentioned produce a region of width about $2^{-41} w(B) = .2 \cdots (10^{-11})$. The discrepancy between this count and the observed 171 may be accounted for by the termination criterion and by the presence of slight increases in the bounding regions from the use of rounded interval arithmetic.

7.3. Nonlinear optimization. An application of interval methods for computing error bounds for an approximate Kuhn–Tucker point of a nonlinear program has been presented by S. M. Robinson (N 3). The method can be applied to a general nonlinear programming problem of the form

(7.9)   minimize $t(x)$ for x in S, an open set in R^n, subject to the constraints: $g(x) \leq 0$, $h(x) = 0$ where t, g, and h are differentiable functions from S into R, R^m, R^q (respectively).

The functions t, g, and h are not necessarily convex. A numerical "solution" of (7.9) is usually a triple $(\bar{x}, \bar{u}, \bar{v})$ with $\bar{x} \in S$, $\bar{u} \in R^m$, $\bar{v} \in R^q$ at which the "first order Kuhn–Tucker conditions"

(7.10)

$$t'(x) + u^T g'(x) + v^T h'(x) = 0$$

$$g(x) \leq 0, \qquad u^T g(x) = 0, \qquad h(x) = 0, \qquad u \geq 0$$

are *approximately* satisfied. Let $\bar{z} = (\bar{x}, \bar{u}, \bar{v})$ be the $p = n + m + q$ dimensional real vector representing the approximate Kuhn–Tucker point. We can enclose $\bar{z}$ in a p-cube by forming $Z = \bar{z} + E$, where E is a p-dimensional interval vector with $m(E) = 0$ and $w(E_i) = \bar{w}$ for some $\bar{w} > 0$ for all $i = 1, 2, \cdots, p$. Thus, $Z_i = \bar{z}_i + [-1, 1](\bar{w}/2)$. Robinson (N 3) gives computable sufficient conditions for the existence of a unique point z in Z which satisfies (7.10) *exactly*.

In order to describe the method of Robinson, we need the following additional notation. Let f be the mapping from R^p into R^p defined for $z = (x, u, v)$, $x \in R^n$, $u \in R^m$, $v \in R^q$, $p = n + m + q$, by

$$(7.11) \quad f_i(z) = \begin{cases} (t'(x) + u^T g'(x) + v^T h'(x))_i^T, & i = 1, \cdots, n, \\ u_{i-n} g_{i-n}(x), & i = n+1, \cdots, n+m, \\ h_{i-n-m}(x), & i = n+m+1, \cdots, n+m+q = p. \end{cases}$$

For any interval $C = [\underline{C}, \bar{C}]$, define

$$C_0 = \begin{cases} [0, \bar{C}] & \text{if } 0 \leq \underline{C}, \\ C & \text{otherwise}; \end{cases}$$

(7.12)

$$C^0 = \begin{cases} [\underline{C}, 0] & \text{if } \bar{C} \leq 0, \\ C & \text{otherwise}. \end{cases}$$

Recall that strict complementary slackness is said to hold at a Kuhn–Tucker point $z = (x, u, v)$ if exactly one of the quantities u_i, $g_i(x)$ is zero for each $i = 1, \cdots, m$. Suppose, for $i, j = 1, 2, \cdots, p$, we have inclusion monotonic interval extensions $F'_{ij}(z)$ of $(\partial f_i/\partial f_j)(z)$. We assume that t, g, and h are twice continuously differentiable in S. Define the interval Newton operator $\bar{N}(z, Z)$, for $z \in Z = (X, U, V)$, $x \in X$, $u \in U$, $v \in V$, by

(7.13) $\bar{N}(z, Z) = z - Mf(z)$ where M is an interval matrix such that $M_r \in M$ for every real matrix M_r which is the inverse of some real matrix contained in the interval matrix $\bar{F}'(z, Z)$ with components $\bar{F}'_{ij}(z, Z) = (\partial f_i/\partial z_j)(z_1, \cdots, z_{j-1}. Z_j, \cdots, Z_p)$.

Robinson proves the following (N 3).

THEOREM 7.3. *Let G be an inclusion montonic interval extension of g and suppose that $0 \notin G_i(Z)_0 \cap U_i^0$ for $1 \leq i \leq m$. Suppose further that $\bar{F}'(Z, Z)$ contains no singular real matrix and that $\bar{N}(\bar{z}, Z) \subseteq Z$. Then $\bar{N}(\bar{z}, Z)$ contains a Kuhn–Tucker triple $z = (x, u, v)$ of (7.9) at which strict complementary slackness holds with linear independence of the gradients to the active constraints; furthermore, z is the only Kuhn–Tucker triple of (7.9) in Z.*

The computationally verifiable condition, $\bar{N}(\bar{z}, Z) \subseteq Z$, in Theorem 7.3 has been shown (N 4) to imply the existence of a unique solution z in Z to $f(z) = 0$. This condition, which requires the computation of an interval matrix containing the set of inverses of real matrices in the interval matrix $\bar{F}'(\bar{z}, Z)$, can be replaced by the simpler condition $K(Z) \subseteq Z$, discussed at length in § 5.2, which only requires an approximate inverse Y of the *real* matrix $f'(\bar{z})$, (or of the real matrix $m(F'(Z))$). The condition $\|I - YF'(Z)\| < 1$ (which follows automatically from $K(Z) \subseteq Z$ for a p-cube Z, (N 6)) implies the nonsingularity of real matrices contained in $F'(Z)$.

Some progress has been made using interval methods, toward the solution of problems of *global* optimization in general nonlinear programming (N 7). The cyclic bisection procedure (4.30) in § 4.4 can find the global minimum of a function in an n-dimensional rectangle.

NOTES

1. See P. D. Roberts and A. Ben-Israel [83], B. Machost [49], N. F. Stewart [92], R. Krawczyk [70, pp. 215–222].

2. R. Dussel [70, pp. 169–177], [13].

3. S. M. Robinson [84].

4. See K. Nickel [67]. The operator $\bar{N}(z, Z)$ was first considered by E. Hansen [24].

5. R. E. Moore [61], [63].

6. R. E. Moore [63].

7. See L. J. Mancini [51], L. J. Mancini and G. P. McCormick [52], R. E. Moore [60], and E. R. Hansen [29].

Chapter 8

Applications to Operator Equations

8.1. Initial value problems in ordinary differential equations. In this section we will consider some applications of interval methods to the construction of *global error bounds* (for all $t_0 \leq t \leq t_1$) along with approximate solutions of systems of ordinary differential equations of the general form

$$(8.1) \qquad x_i'(t) = f_i(t, x_1, x_2, \cdots, x_n), \qquad i = 1, \cdots, n,$$

with given initial values $x_i(t_0)$, $1 \leq i \leq n$. We may write (8.1) in vector notation as $x' = f(t, x)$.

In § 5.3 we illustrated an application of Theorem 5.7 to a particular initial value problem (5.29). The same type of application can be made to practically *any* system of the form (8.1). All we need is inclusion monotonic interval extensions of the functions f_i in the system (8.1). We have discussed a large class $FC_n(X_0)$ of functions f_i *for which this can be done easily* in § 4.4. If $F_i(t, X_1, X_2, \cdots, X_n)$ is an inclusion monotonic, Lipschitz interval extension of $f_i(t, x_1, \cdots, x_n)$ defined for $t_0 \leq t \leq t_1$ and $X_j \subseteq (X_0)_j (i, j = 1, \cdots, n)$, then consider the operator P given by

$$(8.2) \qquad \begin{aligned} &P(X)_i(t) = x_i(t_0) + \int_{t_0}^t F_i(s, X_1(s), \cdots, X_n(s))\, ds, \\ &P(X)(t) = (P(X)_1(t), \cdots, P(X)_n(t)). \end{aligned}$$

In vector notation, we have

$$(8.3) \qquad P(X)(t) = x(t_0) + \int_{t_0}^t F(s, X(s))\, ds.$$

If $w(F(s, X(s))) \leq Lw(X(s))$ for all $t_0 \leq s \leq t_1$, then

$$(8.4) \qquad \sup_t w(P(X)(t)) \leq L(t_1 - t_0) \sup_t w(X(t)), \qquad t_0 \leq t \leq t_1,$$

and the condition (5.28) of Theorem 5.7 is satisfied, for $t_0 \leq t \leq t_1$, if

93

$c = L(t_1 - t_0) < 1$. It follows from Theorem 5.7 that, under these conditions, (8.1) will have a unique solution contained in $X(t)$ for $t_0 \le t \le t_1$ if $P(X)(t) \subseteq X(t)$ for $t_0 \le t \le t_1$. We can *always* find such an $X(t)$ for small enough $t_1 - t_0$, namely *any* constant interval vector function

$$(8.5) \qquad X(t) \equiv B \quad \text{such that} \quad \underline{B} < x_i(t_0) < \bar{B}_i \quad \text{for } i = 1, \cdots, n.$$

For any B satisfying (8.5) we will have $P(B)(t) \subseteq B$ for small enough $t_1 - t_0$. The solution will also be contained in $P(B)(t)$ for $t_0 \le t \le t_1$ (see Theorem 5.7).

We can use any of a variety of methods to improve the bounds on the solution in $[t_0, t_1]$; for instance we can make use of the methods of §§ 3.4 and 3.5 (as we will show in examples to follow). Furthermore, having found a satisfactory approximate solution $\bar{x}(t)$ and global error bounds for all t in $[t_0, t_1]$, say $x(t) - \bar{x}(t) \in E(t)$, we can continue the solution beyond t_1 by replacing $x(t_0)$ in (8.3) by $\bar{x}(t_1) + E(t_1)$ and by repeating the entire procedure, in order to find an approximation and error bounds over a new interval $[t_1, t_2]$. We can continue in a step-by-step fashion.

We will illustrate the method just outlined as applied to an initial value problem discussed by W. F. Ames and E. Adams (N 1), namely:

$$x_1' = a_1 x_1 (1 - x_2),$$
$$(8.6) \qquad x_2' = -a_2 x_2 (1 - x_1),$$
$$x_1(0), x_2(0) \text{ given.}$$

The initial value problem (8.6) is Volterra's model of conflicting populations (N 1). To make things definite, we consider the same numerical coefficients discussed by Ames and Adams. We take $a_1 = 2$, $a_2 = 1$, $x_1(0) = 1$, $x_2(0) = 3$. For the interval extensions we can take $F_1(t, X_1, X_2) = a_1 X_1 (1 - X_2)$, $F_2(t, X_1, X_2) = -a_2 X_2 (1 - X_1)$. In this example, as is the case for any autonomous system, the functions do not depend explicitly on the independent variable t. This is not essential, but does simplify things a little. For B in (8.5) we can take $B_1 = [0, w_1]$ and $B_2 = [0, w_2]$ for any $w_1 > 1$, $w_2 > 3$. Clearly, F_1 and F_2 are inclusion monotonic and Lipschitz for $X \subseteq B$.

We will have $P(B)(t) \subseteq B$ for all $t_0 = 0 \le t \le t_1$ if

$$1 + 2t[1 - w_2, 1]w_1 \subseteq [0, w_1],$$
$$(8.7)$$
$$3 - t[1 - w_1, 1]w_2 \subseteq [0, w_2]$$

for all $0 \le t \le t_1$. This will be the case, for instance, for $w_1 = \frac{4}{3}$, $w_2 = 4$, $t_1 = \frac{1}{8} = 0.125$.

For these numbers, it is not hard to show that we can take $c = \frac{42}{48} < 1$ in (5.28), thus guaranteeing the existence of a unique solution of (8.6) for the given initial conditions and constants for t in $[0, 0.125]$. For the solution $x(t)$, we will have for $0 \le t \le 0.125$,

(8.8)
$$x_1(t) \in 1 + [-8, \tfrac{8}{3}] \, t,$$
$$x_2(t) \in 3 + [-4, \tfrac{4}{3}] \, t.$$

If we take $\bar{x}_1(t) = 1 - \tfrac{8}{3} t$ and $\bar{x}_2(t) = 3 - \tfrac{4}{3} t$, then $x_1(t) - \bar{x}_1(t) \in E_1(t) = \tfrac{16}{3}[-1, 1] \, t$ and $x_2(t) - \bar{x}_2(t) \in \tfrac{8}{3}[-1, 1] \, t$. Put another way, $|x_1(t) - \bar{x}_1(t)| \le \tfrac{16}{3} \, t$ and $|x_2(t) - \bar{x}_2(t)| \le \tfrac{8}{3} \, t$. So far these are very crude bounds on the solution. We can obtain *arbitrarily sharp bounds* as follows.

From the crude bounds (8.8) we can determine that, in $0 \le t \le 0.125$, we have $x_1(t) \in [0, \tfrac{4}{3}]$ and $x_2(t) \in [2.5, \tfrac{19}{6}]$. We can use these interval bounds to bound the remainder in Taylor series expansions of the solution as follows. We refer to §3.4 for the derivation of recursion formulas for the successive Taylor coefficients in the form:

(8.9)
$$T_1 = 1 - x_2, \qquad (T_1)_k = -(x_2)_k \qquad (k \ge 1),$$

$$T_2 = x_1 \, T_1, \qquad (T_2)_k = \sum_{j=0}^{k} (x_1)_j (T_1)_{k-j},$$

$$(x_1)_1 = 2 \, T_2, \qquad (x_1)_{k+1} = (2/(k+1))(T_2)_k,$$

$$T_3 = 1 - x_1, \qquad (T_3)_k = -(x_1)_k,$$

$$T_4 = x_2 \, T_3, \qquad (T_4)_k = \sum_{j=0}^{k} (x_2)_j (T_3)_{k-j},$$

$$(x_2)_1 = -T_4, \qquad (x_2)_{k+1} = (-1/(k+1))(T_4)_k.$$

The recursion formulas (8.9) can be evaluated either in ordinary machine arithmetic or in rounded interval arithmetic. If we put $X_1 = [0, \tfrac{4}{3}]$ and $X_2 = [2.5, \tfrac{19}{6}]$ we can evaluate the formulas (8.9) in rounded interval arithmetic to find intervals $R_1^{(k)}$ and $R_2^{(k)}$ containing, respectively, the range of values of the kth Taylor coefficients of $x_1(t)$ and $x_2(t)$ for all t in $[0, 0.125]$. We can also evaluate the recursion formulas (8.9) in ordinary machine arithmetic, or in rounded interval arithmetic (in order to bound round-off error) using the given initial values $x_1(0) = 1$ and $x_2(0) = 3$. We obtain, in this way, interval polynomial enclosures of the solution of the form (here, $t_0 = 0$)

(8.10)
$$x_1(t) \in X_1^{(N)}(t) = \sum_{k=0}^{N-1} (x_1)_k (t - t_0)^k + R_1^{(N)}(t - t_0)^N,$$

$$x_2(t) \in X_2^{(N)}(t) = \sum_{k=0}^{N-1} (x_2)_k (t - t_0)^k + R_2^{(N)}(t - t_0)^N.$$

We can take $\bar{x}(t) = m(X^{(N)}(t))$ and $E(t) = X^{(N)}(t) - m(X^{(N)}(t))$. The recursion formulas (8.9) are derived only *once* for a particular system of differential equations. They can be used to evaluate the coefficients in Taylor expansions about *any* point. To find the Taylor coefficients $(x_1)_k$ and $(x_2)_k$ for $k = 0, 1, \cdots,$ $N - 1$ in (8.10) we compute and store, a column at a time, the elements in the array, say for $N = 9$, (carried out here in 8 decimal digit real arithmetic).

$$
\begin{array}{llll}
(x_1)_0 = 1 & (x_1)_1 = -4 & (x_1)_2 = 8 & \cdot \quad (x_1)_8 = 127.10159 \cdots \\
(x_2)_0 = 3 & (x_2)_1 = 0 & (x_2)_2 = -6 & \cdot \quad (x_2)_8 = -64.77857 \cdots \\
(T_1)_0 = -2 & (T_1)_1 = 0 & (T_1)_2 = 6 & \cdot \\
(T_2)_0 = -2 & (T_2)_1 = 8 & (T_2)_2 = -10 & \cdot \\
(T_3)_0 = 0 & (T_3)_1 = 4 & \cdots & \cdot \\
(T_4)_0 = 0 & (T_4)_1 = 12 & \cdots & \cdot
\end{array}
$$

(8.11)

If we want to find the intervals $R_1^{(N)}$ and $R_2^{(N)}$ we can evaluate the same recursion formulas (8.9) in interval arithmetic beginning with $(X_1)_0 = [0, \frac{4}{3}]$ and $(X_2)_0 = [2.5, \frac{19}{6}]$. It is convenient, for scaling purposes, to compute the quantities $R_1^{(N)}(t_1 - t_0)^N$ and $R_2^{(N)}(t_1 - t_0)^N$ in (8.10) directly. To do this we need only modify the recursion formulas slightly, merely multiplying the expressions for $(x_1)_{k+1}$ and $(x_2)_{k+1}$ by $(t_1 - t_0)$. In this way we will obtain the quantities $(\cdot)_k(t_1 - t_0)^k$ as we fill out the array. We obtain, for instance,

(8.12)
$$
R_1^{(9)} .125^9 \subseteq [-00007556, .00005833],
$$
$$
R_2^{(9)} .125^9 \subseteq [-.00003125, .00004042].
$$

To find $R_1^{(9)}(t - t_0)^9$ we can evaluate $((t - t_0)/.125)^9[-.00003125, .00004042]$. For instance $R_1^{(9)} .05^9 \subseteq [-.0000\,000082, .0000\,00011]$.

Using 8 decimal digit rounded interval arithmetic in (8.10) we obtained the following intervals for the evaluation of $X_2^{(N)}(t)$ at $t = 0.125$.

N	$X_2^{(N)}(0.125)$	$\frac{1}{2}w(X_2^{(N)}(0.125))$
1	$[2.60\ldots\ldots, 3.13\ldots\ldots]$	$.265\ldots$
2	$[2.8488\ldots, 3.0247\ldots]$	$.0879\ldots$
3	$[2.89727\ldots, 2.95898\ldots]$	$.0308\ldots$
4	$[2.912873\ldots, 2.93087\ldots]$	$.00899\ldots$
5	$[2.919739\ldots, 2.92449\ldots]$	$.00237\ldots$
6	$[2.920705\ldots, 2.92287\ldots]$	$.00108\ldots$
7	$[2.921161\ldots, 2.921854\ldots]$	$.00034\ldots$
8	$[2.921444\ldots, 2.921669\ldots]$	$.00011\ldots$
9	$[2.9215249, 2.9215965]$	$.000035\ldots$

By carrying enough precision in the rounded interval arithmetic and by using any $N \geq 1$, we can compute arbitrarily sharp bounds on the values of the exact solution for $0 \leq t \leq 0.125$, by subdividing the interval $[0, t_1] = [0, .125]$. Let $t^{(j)} = jh$, where $h = .125/M$ for some integer $M > 1$. Thus $t^{(0)} = 0$ and $t^{(M)} = .125$. We do not need to recompute the quantities $R_1^{(N)}$ or $R_2^{(N)}$ since these bound the Nth terms in the Taylor expansion at *any* t in $[0, .125]$. If we recompute the coefficients $(x_1)_k$, $(x_2)_k$ for $k = 0, 1, \cdots, N - 1$ using the initial values at $t = 0$ and using $X_1^{(N)}(t^{(j)})$, $X_2^{(N)}(t^{(j)})$, from (8.10), at $t^{(j)}$ for $j = 1, 2, \cdots, M - 1$, then we can obtain sharper bounds on the solution over the interval $[0, .125]$ than those provided by a single expansion of the form (8.10). In fact the width of the resulting interval solution, using subdivisions, will be decreased by a factor approximately equal to $(1/M)^N$ down to the point where roundoff error "takes over". For a fixed value of N, we could *compute* and *intersect* the interval values of $X_1(t)$ obtained for an increasing sequence $\{M_g\}_q = 1, 2, \cdots$ evaluated at any particular t in $[0, .125]$. The process would converge in a finite number of steps in rounded interval arithmetic carried out in fixed precision. Sharper bounds would then require higher precision arithmetic.

We can also apply the methods of this section to computing bounds on the *set* of solutions to an initial value problem resulting from intervals of possible values for constants in the differential equations and intervals of possible initial values. The coefficients $(x_1)_k$, $(x_2)_k$, $\cdots$, $(x_n)_k$ will, in such a case, be intervals from the outset. The condition (8.5) can be replaced by

$$(8.13) \qquad X(t) \equiv B \quad \text{such that} \quad \underline{B} < \underline{X}_i(t_0) \leq \bar{X}_i(t_0) < \bar{B}_i \quad \text{for } i = 1, 2, \cdots, n.$$

The subdivision-intersection technique described in the previous paragraph will, in this case, produce a sequence of intervals converging in a finite number of steps to an interval containing the entire set of values of solutions at a particular value of t. For example, we can *continue* the interval solution of the problem (8.6) from the new, *interval* initial conditions $X_1^{(N)}(.125)$, $X_2^{(N)}(.125)$.

Numerous additional studies of applications of interval methods to initial value problems appear in the literature (N 2). An application of interval analytic methods to the determination of *periodic solutions* of system of ordinary differential equations has been discussed by U. Waushkuhn (N 3).

A difficulty arises in connection with the continuation of sets of solutions to an initial value problem for a system of n ordinary differential equations. Suppose a set of initial points $x(0)$ is enclosed in a finitely representable set S_0 such as an interval vector, an ellipsoid, or a polytope. The set of solution points

$$S_t = \{x(t) = (x_1(t), \cdots, x_n(t)): x(0) \in S_0\}$$

for $t > 0$ will, in general, not be exactly representable by the same type of geometrical object as S_0. Thus, if S_{t_1} is enclosed again in an interval vector, ellipsoid, or polytope, the resulting bounding region will contain extra points, not in the solutions emanating from S_0. This phenomenon has been dubbed *"the wrapping effect"* and has been studied in a variety of ways (N 24). Successful attempts to reduce the growth of bounds due to the wrapping effect have been based on various ideas including the following:

1. We can represent a bounding region by a triple (y^*, M, Z) consisting of an

approximate solution y^*, a matrix (linear transformation) M, and an interval vector Z—all time dependent—so that

$$S_t \subseteq y^*(t) + M(t)Z(t).$$

This gives us the flexibility of carrying out the bounding computations in a moving, nonrectangular coordinate system.

2. We can sharpen (but complicate) the representation of a bounding region containing S_t at appropriate values of t by adding new faces to bounding polytopes or new elements to finite coverings of S_t with unions of bounding regions.

3. We can carry out analytic changes of variables before or during the computation to reformulate the differential equations in more convenient variables so that the direction field is, to some extent, straightened out—leading to less distortion of the types of bounding regions chosen (N 25).

4. We can use nonrectangular norms, for instance ellipsoidal norms.

5. We can make use of linearities and monotonicities for some systems of differential equations to construct bounding regions from finite sets of particular approximate solutions (N 23).

6. G. W. Harrison has recently found a way to eliminate the wrapping effect when solving a system of linear differential equations with interval valued coefficients (N 27):

$$x' = Ax + r, \qquad x(0) \text{ given,}$$

$$\text{with } \underline{A}_{ij} \leqq A_{ij} \leqq \bar{A}_{ij} \text{ and } \underline{r}_i \leqq r_i \leqq \bar{r}_i$$

$$i, j = 1, 2, \cdots, n.$$

For the applications of interest to Harrison (linear compartmental models of biological phenomena such as *pharmacokinetics* or *nutrient cycling in ecosystems*), the diagonal elements of A are given by

$$A_{ii} = -A_{0i} - \sum_{k \neq i} A_{ki}$$

where $A_{0i} x_i$ is the "flow rate from compartment i to outside the system". Harrison proves the following.

Let p^T denote the transpose of a vector p. Let e_i denote the ith column of the $n \times n$ identity matrix. Choose $t_i > 0$. Put $p(t_1) = e_i$ and integrate

$$dp^T/dt = -p^T A^*(t)$$

backwards from $t = t_1$ to $t = 0$ to get $p^T(0)$ and $p^T(0)x(0)$. Then integrate

$$d(p^T \bar{x})/dt = p^T r^* \quad \text{with } \bar{x}(0) \geqq x(0)$$

forward from $t = 0$ to $t = t_1$ to get $(p^T \bar{x})(t_1) = \bar{x}_i(t_1)$ with the result that $x_1(t_1) \leq \bar{x}_i(t_1)$ where A^* and r^* are given by

$$A_{ij}^*(t) = \begin{cases} \bar{A}_{ij} & \text{if } p_i(t) - p_j(t) \geq 0, \\ \underline{A}_{ij} & \text{if } p_i(t) - p_j(t) < 0 \end{cases}$$

for $i \neq j$ and $i \neq 0$ and

$$A_{0i}^*(t) = \begin{cases} \underline{A}_{0i} & \text{if } p_i(t) \geq 0, \\ \bar{A}_{0i} & \text{if } p_i(t) < 0 \end{cases}$$

and

$$A_{ii}^*(t) = -A_{0i}^*(t) - \sum_{k \neq i} A_{ki}^*(t)$$

and

$$r_i^*(t) = \begin{cases} \bar{r}_i & \text{if } p_i(t) \geq 0, \\ \underline{r}_i & \text{if } p_i(t) < 0. \end{cases}$$

The same procedure, repeated with $p(t_1) = -e_i$ yields $\underline{x}_i(t_1)$ such that $\underline{x}_i(t_1) \leq x_i(t)$. It is shown (N 27) that the bounds thus computed are sharp. It remains to investigate the effects of roundoff and truncation error in approximate solutions of the required equations.

A number of applications to initial value problems have been made of successive Taylor series expansions with *recursive generation of Taylor coefficients* (see § 3.4). We can carry out such a method in ordinary machine arithmetic, omitting the remainder term, to obtain an approximate numerical solution; or we can carry out such a method in interval arithmetic including the remainder term to obtain rigorous upper and lower bounds on the solution. Recursion relations for successive Taylor coefficients in the *n-body problem in Newtonian gravitational theory* have been given (N 29) and used to reconfirm a prediction of the theory of relativity concerning the motion of the perihelion of Mercury over the real time period 1850–1950. Using 12th order Taylor series expansions from step to step, we obtained excellent agreement with a previous computation of the Newtonian prediction for the motion of Mercury, which differs from the observed motion by almost exactly what is predicted by relativity.

D. Greenspan (N 28) has used 8th order successive Taylor series expansions with recursive generation of Taylor coefficients to find, approximately, *periodic solutions of van der Pol's equation.*

Recursive generation of Taylor coefficients is also used in eigenfunction expansions of solutions of *linear, separable partial differential equations* on rectangular regions by Y. F. Chang, G. Corliss, and G. Kriegsman (N 30).

8.2. Two-point boundary value problems and integral equations. A two point
boundary value problem of the form

(8.14) $y'' = f(t, y), \qquad y(0) = y(1) = 0,$

can be written in the form of an integral equation

(8.15) $y(t) = (t-1) \int_0^t sf(s, y(s)) \, ds + t \int_t^1 (s-1)f(s, y(s)) \, ds.$

We can regard the right hand side of (8.15) as defining an operator p so that
(8.15) is an operator equation of the form $y(t) = p(y)(t)$, discussed in § 5.3. If f
is in $FC_n(X_0)$ (see § 4.4), then it has an inclusion monotonic interval extension
F. The interval operator P defined by

(8.16) $P(Y)(t) = (t-1) \int_0^t sF(s, Y(s)) \, ds + t \int_t^1 (s-1)F(s, Y(s)) \, ds$

is an inclusion monotonic interval majorant of p; see § 5.3. Suppose that
$P(Y^{(0)})$ is defined for some continuous interval function $Y^{(0)}(t)$ for which
$Y^{(0)}(0)$ and $Y^{(0)}(1)$ both contain the boundary value 0. Let $B \supseteq F([0, 1],$
$Y^{(0)}([0, 1]))$. We have

$$P(Y^{(0)})(t) \subseteq (t-1)(t^2/2)B - t((t-1)^2/2)B$$

(8.17) $\subseteq ((t-1)t/2)(tB + (1-t)B)$

$$\subseteq ((t-1)t/2)B \quad \text{for } 0 \le t \le 1.$$

Thus, if $((t-1)t/2) B \subseteq Y^{(0)}(t)$ for all t in $[0, 1]$, then we will have $P(Y^{(0)}) \subseteq Y^{(0)}$
and Theorem 5.7 applies, at least through property (3). If (5.28) holds, then
(8.14) has a unique solution in $P(Y^{(0)})(t)$ given by (5.27).

We find, for example, that the two-point boundary value problem

(8.18) $y'' = \frac{3}{2}(y + 1 + 1.126t)^2 + .625t, \qquad y(0) = y(1) = 0,$

has a unique solution in $P(Y^{(0)})(t) \subseteq [-1.3755, -.2112](1-t) \, t$ where $Y^{(0)}$ was
chosen as the constant interval $[-.35, 0]$.

Similarly, applying the same techniques, we find that the two-point boundary
value problem

(8.19) $y'' = e^{-y}, \qquad y(0) = y(1) = 0$

has a unique solution *in* $[-.582, -.5](1-t) \, t$. Here, $Y^{(0)}$ was chosen as $[-.15,$
0], (N 4). This particular problem does have another solution; however, it is
not contained in the interval function shown, but goes outside to more negative
values.

Theorem 5.7 provides a basis via (5.26) for the computation of arbitrarily
sharp bounds on solutions of (8.15). Alternatively, for nonlinear problems, we

could use the crude interval bounds as obtained above to select a starting approximation for an iterative noninterval method such as a finite difference method, a collocation method, or the ordinary Picard iteration method. In this latter connection, see Theorem 5.8. For instance, we can take $y^{(0)}(t) = m(P(Y^{(0)}(t)))$.

The techniques of § 3.5 are useful in interval iteration methods for improving bounds on solutions of (8.15) using (5.26).

The same methods can be used to compute bounds on the *set* of solutions to two-point boundary value problems and integral equations when constants in the equations are allowed to vary over intervals of possible values. For example, the problem

$$(8.20) \qquad y'' = (1 + ct^2)y + e^{at}, \qquad y(0) = y(1) = 0$$

when $.9 \leq c \leq 1.1$ and $.8 \leq a \leq .9$ has solutions *all of which* can be shown, by the methods just described, to lie in the interval function $[-1.65, -.038](1-t)t$ for $0 \leq t \leq 1$.

To do this, we put $F(s, Y(s)) = A(s) + B(s)Y(s)$ in (8.16) where

$$e^{as} \in A(s) \quad \text{for all } s \text{ in } [0, 1] \text{ and all } a \text{ in } [.8, .9],$$

$$1 + cs^2 \in B(s) \quad \text{for all } s \text{ in } [0, 1] \text{ and all } c \text{ in } [.9, 1.1].$$

We can take $B(s) = 1 + [.9, 1.1]s$ and

$$A(s) = 1 + A_0 s + \cdots + (A_0 s)^q/q! + A_1(A_0 s)^{q+1}/(q+1)!$$

for any nonnegative integer q, where $A_0 = [.8, .9]$ and A_1 is an interval containing $[1, e^{.9}]$, say $A_1 = [1, 2.5]$.

It is not hard to show that, for any q, we have $A(s) \subseteq [1, 3.3]$ for all s in $[0, 1]$ and that $B(s) \in [1, 2.1]$. Thus, we will have $P(Y^{(0)}) \subseteq Y^{(0)}$ for any q if it is true using $A(s) \equiv [1, 3.3]$ and $B(s) \equiv [1, 2.1]$.

If we put $Y^{(0)}(t) \equiv [-.44, 0]$, for instance, then

$$P(Y^{(0)})(t) \subseteq \{[1, 3.3] + [1, 2.1][-.44, 0]\}(t-1)t/2$$

for all t in $[0, 1]$. From this, it follows that $P(Y^{(0)}) \subseteq Y^{(0)}$ and so the interval function $[-1.65, -.038](1-t)t$ contains the set of all solutions to (8.20) for c in $[.9, 1.1]$ and a in $[.8, .9]$.

The more general two-point boundary value problem

$$(8.21) \qquad \begin{array}{l} y'' = f(t, y, y'), \\[6pt] g_1(a, y(a), y'(a)) = 0, \qquad g_2(b, y(b), y'(b)) = 0 \end{array}$$

is discussed by E. Hansen (N 6). By subdividing the interval $[a, b]$ and using finite difference methods, with error terms evaluated in interval arithmetic, Hansen obtains arbitrarily sharp bounds on the solutions of problems of type (8.21).

An interval version of the Newton–Kantorovich method in function spaces is discussed in the literature as well as other interval methods for two-point boundary value problems (N 7). The use of methods from *functional analysis* for the numerical solution of operator equations can be aided by interval computation of bounds on norms of functions, norms of operators, etc. This usually involves bounding the range of values of various functions or sets of functions and can be carried out using the methods described in Chapters 3, 4, and 5. This can be done both to verify sufficient conditions for existence and convergence (for instance the so-called "Kantorovich conditions") as well as to compute numerical bounds on the error in an approximate solution.

An interval method for certain Fredholm integral equations has been given by W. Neuland (N 8).

8.3. Partial differential equations. Error bounds for an approximate solution of a class of elliptic partial differential equations have been obtained with the help of interval methods by W. Appelt (N 9). The results can be described as follows.

Let G be a *normal domain*: an open, simply connected, bounded domain in E^2 in which the Gauss integral theorem holds. Let $\bar{G} = G \cup \partial G$, where ∂G is the boundary of G.

Consider elliptic differential operators D on $C^2[G]$ of the form

(8.22) $$Du = a_{20}u_{xx} + a_{11}u_{xy} + a_{02}u_{yy} + a_1u_x + a_2u_y,$$

where

$$a_{20}, a_{11}, a_{02}, a_1, a_2 \in C[\bar{G}];$$

and having the following property (uniformly elliptic in G):

there is a constant $m > 0$ such that, for all (x,y) in $\bar{G}$ and all (ξ_1, ξ_2) in E^2,

$$a_{20}(x,y)\xi_1^2 + a_{11}(x,y)\xi_1\xi_2 + a_{02}(x,y)\xi_2^2 \geq m(\xi_1^2 + \xi_2^2)$$

holds.

(The canonical example of such an operator is the Laplacian $u_{xx} + u_{yy}$.)

Given f in $C(\bar{G})$ and g in $C(\partial G)$, consider the boundary value problem

(8.23)
$$Du = f \quad \text{in } G,$$
$$u = g \quad \text{on } \partial G.$$

A function u in $C(\bar{G}) \cap C^2(G)$ which satisfies the boundary value problem is called a solution.

For elliptic operators of the type considered, the *minimum-maximum principle* holds:

Let u be a nonconstant function in $C[\bar{G}]$, then

(1) $Du \leq 0$ in G implies that the *minimum value* of u occurs on ∂G; and in G, $u(x,y)$ is strictly greater than this;

(2) $Du \geqq 0$ in G implies that the *maximum value* of u occurs on ∂G; and in G, $u(x, y)$ is strictly less than this.

It follows that, if the boundary value problem has a solution, it is unique.

If an approximate solution $\bar{u}$ to the boundary value problem can be found for which $\bar{u}$ is in $C[G] \cap C^2[G]$ and

$$D\bar{u} = \bar{f} \quad \text{in } G,$$

$$\bar{u} = \bar{g} \quad \text{on } \partial G$$

then the *residual* or *defect* of $\bar{u}$ is defined to be

$$g - \bar{g} \text{ on } \partial G \quad \text{and} \quad f - \bar{f} \text{ in } G.$$

Suppose Q is another normal domain in E^2 for which $G \subseteq Q$ and suppose t is in $C[\bar{Q}] \cap C^2[Q]$ and

$$Dt \leqq -1 \quad \text{in } Q,$$

$$t \geqq 0 \quad \text{on } \partial Q$$

then, for all (x,y) in $\bar{G}$, it follows from the maximum principle that

(8.24) $$|u - \bar{u}| \leqq \max_{\partial G} |g - \bar{g}| + t(x,y) \max_{\bar{G}} |f - \bar{f}|.$$

Such functions t can be found for a variety of domains G including any that can be enclosed in a domain Q whose boundary is a circle, rectangle, or a sector of a circular ring. The sharpness of the bound will depend on the size of the enclosing region $\bar{Q}$. For Laplace's equation on the unit circle $G = \{(x,y)|x^2 + y^2 < 1\}$, $t(x,y) = \frac{1}{4}(1 - x^2 - y^2)$ satisfies the requirements for the defect procedure on $\bar{Q} = \bar{G}$. Some discussion of optimality criteria for choosing t and Q is given in the paper of Appelt (N 9).

A method for the construction of a twice continuously differentiable approximate solution u to the boundary value problem is based on a bicubic spline function interpolation of a discrete solution obtained by some finite difference method. In order to take into account rounding errors in the computation of the spline function coefficients, interval arithmetic is used so that, finally, using *interval spline functions*, bounds are obtained for

$$a = \max_{\bar{G}} |f - \bar{f}| \quad \text{and} \quad b = \max_{\partial G} |g - \bar{g}|$$

where $\bar{f} = D\bar{u}$ and $\bar{g} = \bar{u}$ on ∂G. These computed bounds are then used in the defect procedure which yields

(8.25) $$|u(x,y) - \bar{u}(x,y)| \leqq b + t(x,y)a \quad \text{in } \bar{G}.$$

The interval spline function $S(x,y)$ that is to be used to contain a real valued spline $s(x,y)$ (in order to cover the effects of rounding errors in computing the coefficients of $s(x,y)$) must be *derivative compatible* (see § 4.5 and Note 13 in Chapter 4); that is, the formal derivatives of S must contain the corresponding derivatives of s. This will be the case if, in $[x_i, x_{i+1}] \times [y_j, y_{j+1}]$, we have

$$(8.26) \qquad s(x,y) = s_{ij}(x,y) = \sum_{m+n \leq 3} a^{ij}_{n,m} (x - x_i)^m (y - y_j)^n$$

and

$$(8.27) \qquad S(x,y) = S_{ij}(x,y) = \sum_{m+n \leq 3} A^{ij}_{n,m} (x - x_i)^m (y - y_j)^n$$

with

$$a^{ij}_{n,m} \in A^{ij}_{n,m}.$$

Numerical results are given by Appelt for five examples including the following two. The programs were written in FORTRAN IV and carried out using double precision (16.8 decimals) on the IBM/370-165. Storage requirements were some 250K using overlay-structure.

Example 1.

$$(8.28) \qquad \begin{aligned} Du &= u_{xx} + u_{yy} = 0 \quad \text{in } G = [0,1] \times [0,1], \\ u &= x^2 - y^2 + 2 \quad \text{on } \partial G \end{aligned}$$

Using successive overrelaxation, a discrete approximate solution was found with square mesh of size $h = \frac{1}{8}$ (49 interior points). The function t was selected as

$$t(x,y) = \tfrac{1}{4}(x - x^2 + y - y^2).$$

The maximum resulting error bound on $|u - \bar{u}|$ in $\bar{G}$ was about 10^{-9}. Computation time was about 10 seconds.

Example 2. Here, G was taken as the interior of the figure shown:

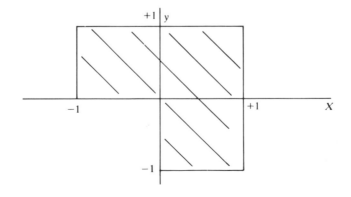

Again, in this example as in the first,

$$Du = u_{xx} + u_{yy} = 0 \quad \text{in } G,$$

$$u = x^2 - y^2 + 2 \quad \text{on } \partial G.$$

Here, t was $t(x,y) = \frac{1}{4}(2 - x^2 - y^2)$. The step size $h = \frac{1}{8}$ was again used resulting in 147 interior points. After interval spline functions were found the resulting error bound on $|u - \bar{u}|$ was less than 10^{-7} in the whole region. Total computation time was about 54 seconds.

Other applications of interval methods to partial differential equations have been made by F. Krückberg and by R. Tost (N 10). Interval methods have been used by E. Adams and W. F. Ames on a nonlinear parabolic partial differential equation which is discretized and then solved iteratively (N 31).

8.4. Monotone operators and differential inequalities. Interval analysis provides more general methods for bounding solutions of operator equations than some of the methods based on differential inequalities, isotone and antitone operators, etc. (N 11). For example, Theorem 5.7 in § 5.3 includes, as special cases, a number of such methods—as we will show in this section. We will also discuss some other applications of interval methods *in combination* with results from the theory of differential inqualities.

A set M_r of real vector valued functions with common domain D can be partially ordered by

(8.29) $x \leqq y$ if and only if $x_i(t) \leqq y_i(t)$ for all $t \in D$, $i = 1, \cdots, n$.

Now suppose that u and v are in M_r and that $u \leqq v$. We define

(8.30) $[u, v] := \{x : x \in M_r, u \leqq x, x \leqq v\}.$

Thus, $[u, v]$ is the *interval* of functions in M_r between u and v.

For each t in D we denote, by $[u(t), v(t)]$, the interval vector $([u_1(t), v_1(t)],$ $\cdots, \lfloor u_n(t), v_n(t) \rfloor])$. Thus, $[u, v]$ is an interval vector valued function Y with values $Y(t) = [u(t), v(t)]$. We will not distinguish between u and $[u, u]$. Let M be a set of M_r-interval vector valued functions on D with $M_r \subseteq M$. Thus, for any u and v in M_r with $u \leqq v$, M contains the interval function $[u, v]$.

Now suppose that x, y, u, v are elements of M_r with $x \leqq y$ and $u \leqq v$. We consider the interval functions $Y = [x, y]$ and $V = [u, v]$ in M. If $u \leqq x$ and $y \leqq v$, then we will have $Y \subseteq V$ (see (5.21)).

Consider an operator $T_r : M_r \to M_r$ and suppose that $T : M \to M$ is an inclusion monotonic interval majorant of T_r (see § 5.3). For example, the interval operator defined on the set M of intervals $[u, v]$ with u, v in M_r by

(8.31) $T[u, v] = \{T_r x : u \leqq x \leqq v\}$

is such an operator. In fact, for $u \leqq x \leqq y \leqq v$, with x, y, u, v in M_r, we have

(8.32) $T[x, y] \subseteq T[u, v].$

Now suppose that a *sequence* of intervals is generated by

(8.33) $[u_{n+1}, v_{n+1}] := T[u_n, v_n],$ $n = 0, 1, 2, \cdots,$

with $u_0 \le u_1 \le v_1 \le v_0$. It follows from Theorem 5.7, that the sequence is *nested*:

(8.34) $u_n \le u_{n+1} \le v_{n+1} \le v_n$ for all n.

These concepts from interval analysis include a number of results from the theory of monotone operators as we will now illustrate.
  Example 1 (L. Collatz (N 11)). Let

$$T^*u = T_1 u + T_2 u + r$$

with

$$T_1 \text{ isotone}: u \le v \quad \text{implies} \quad T_1 u \le T_1 v$$

and

$$T_2 \text{ antitone}: u \le v \quad \text{implies} \quad T_2 v \le T_2 u.$$

Define the sequences $\{u_n\}$, $\{v_n\}$ by

(8.35)
$$u_{n+1} = T_1 u_n + T_2 v_n + r,$$
$$v_{n+1} = T_1 v_n + T_2 u_n + r.$$

If $u_0 \le u_1 \le v_1 \le v_0$, then $u_n \le u_{n+1} \le v_{n+1} \le v_n$ for all n.
  This result from the theory of monotone operators also follows from Theorem 5.7, using (8.34), since the sequences (8.35) are both generated simultaneously from (8.33) by the single *interval* operator

$$T[u, v] = T_1[u, v] + T_2[u, v] + r$$

where $T_1[u, v] = [T_1 u, T_1 v]$ and $T_2[u, v] = [T_2 v, T_2 u]$. The operator T is of the form (8.31) with $T_r = T^*$ and does satisfy (8.32).
  Example 2 (W. F. Ames and E. Adams (N 12)). A Volterra model for conflicting populations $p(t)$ and $q(t)$ can be put into the form

$$p' = ap(1-q),$$
$$q' = -cq(1-p)$$

with $a, c > 0$ and $p(0)$, $q(0)$ prescribed.
  Define the sequences $\{p_n\}$, $\{q_n\}$ by

(8.36)
$$p'_{n+1} = ap_{n+1}(1-q_n),$$
$$q'_{n+1} = -cq_n(1-p_n), \qquad t > 0,$$
$$p_n(0) = p(0), \qquad q_n(0) = q(0).$$

Choosing $p_0(t) = 0$, it is found (N 12) that (8.36) is a *bilateral algorithm*, producing the results:

$$p_0 < p_2 < p_4 < \cdots < p_5 < p_3 < p_1,$$

$$q_0 < q_2 < q_4 < \cdots < q_5 < q_3 < q_1.$$

These results also follow as a special case of Theorem 5.7, using (8.34), as we will now show.

We define the real operators f and g by

(8.37)
$$f(p) = q(0) \exp\left[-c\left(t - \int_0^t p(s)\, ds\right)\right],$$

$$g(q) = p(0) \exp\left[a\left(t - \int_0^t q(s)\, ds\right)\right].$$

Clearly, f is isotone and g is antitone:

$$p_1 < p_2 \quad \text{implies} \quad f(p_1) < f(p_2),$$

$$q_1 < q_2 \quad \text{implies} \quad g(q_1) > g(q_2).$$

The composite mappings $fgfg$ and $gfgf$ are both isotone. From (8.36) and (8.37) we have

(8.38)
$$p_{n+2} = g(f(g(f(p_n)))),$$

$$q_{n+2} = f(g(f(g(q_n)))).$$

Now let u and v be the vector valued functions

$$u = \binom{p}{q}, \qquad v = \binom{r}{s}$$

and define the interval operator

(8.39)
$$T[u, v] = \left[\binom{gfgf(p)}{fgfg(q)}, \binom{gfgf(r)}{fgfg(s)}\right].$$

Clearly, in this notation, (8.36) produces the *same sequences*, $\{p_{2n}\}$, $\{q_{2n}\}$, $\{p_{2n+1}\}$, $\{q_{2n+1}\}$ as are produced by (8.33) with T defined by (8.39) if

$$u_n = \binom{p_{2n}}{q_{2n}} \quad \text{and} \quad v_n = \binom{p_{2n+1}}{q_{2n+1}}.$$

Of course, we need u_0 and v_0 to begin with; thus we need p_0, q_0, p_1, and q_1 to start the interval iteration.

The operator T defined by (8.39) is of the form (8.31), where T_r is given by

$$T_r\binom{p}{q} = \binom{gfgf(p)}{fgfg(q)}.$$

It can be verified that

$$T[u_0, v_0] \subseteq [u_0, v_0]$$

since

$$p_0 < p_2 < p_3 < p_1 \quad \text{and} \quad q_0 < q_2 < q_3 < q_1.$$

It follows from Theorem 5.7, using (8.34), that for the sequences defined by (8.36) we have

$$\binom{p_{2n}}{q_{2n}} \le \binom{p_{2n+2}}{q_{2n+2}} \le \binom{p_{2n+3}}{q_{2n+3}} \le \binom{p_{2n+1}}{q_{2n+1}}$$

for all n.

An extensive study of "bilateral algorithms" has been carried out by W. F. Ames and M. Ginsberg (N 19). In general, the sequences of *pairs* of alternating lower and upper bounds produced by such algorithms can also be produced by the interval algorithm (8.33).

Example 3 (G. Scheu and E. Adams (N 13)). The initial value problem

$$y' = f(t, y), \qquad y(t_0) = a$$

is considered for m-dimensional real vector valued functions y and f. Assumptions are made on f so that the problem has a unique solution $y(t)$ for t in $[t_0, T]$. A pair of operators is defined by

(8.40)
$$\bar{V}(u, v)(t) = a + \bar{e} + \int_{t_0}^{t} \max_{u \le y \le v} f(s, y(s)) \, ds,$$

$$\underline{V}(u, v)(t) = a - \underline{e} + \int_{t_0}^{t} \min_{u \le y \le v} f(s, y(s)) \, ds$$

where $\underline{e}$ and $\bar{e}$ are positive vectors.

It is assumed that vector functions $\underline{y}_0$ and $\bar{y}_0$ can be found such that

(8.41)
$$\bar{V}(\underline{y}_0, \bar{y}_0)(t) \le \bar{y}_0(t),$$

$$\underline{y}_0(t) \le \underline{V}(\underline{y}_0, \bar{y}_0)(t)$$

for all t in some interval $[t_0, T]$.

The sequences $\{\underline{y}_n\}$, $\{\bar{y}_n\}$ are then defined by

(8.42)
$$\underline{y}_{n+1}(t) = \underline{V}(\underline{y}_n, \bar{y}_n)(t)$$

$$\bar{y}_{n+1}(t) = \bar{V}(\underline{y}_n, \bar{y}_n)(t), \qquad n = 0, 1, 2, \cdots,$$

and the theory of differential inequalities and quasi-monotone operators is used to show (N 13) that (8.41) implies that

(8.43) $$\underline{y}_0 \le \underline{y}_1 \le \cdots \le \underline{y}_n \le \cdots \le \underline{y} \le y \le \bar{y} \le \cdots \le \bar{y}_n \le \cdots \le \bar{y}_1 \le \bar{y}_0$$

where y and $\bar{y}$ are the limits of the sequences $\{y_n\}$ and $\{\bar{y}_n\}$ respectively and y is the unique solution of the initial value problem.

This result is again a special case of Theorem 5.7 as we now show.

Consider the interval operator defined by

$$(8.44) \qquad T[u, v](t) = a + [-\underline{e}, \bar{e}] + \int_{t_0}^{t} F(s, [u(s), v(s)]) \, ds$$

where

$$(8.45) \qquad F(s, [u(s), v(s)]) \supseteq [\min_{u \leq y \leq v} f(s, y(s)), \max_{u \leq y \leq v} f(s, y(s))].$$

Here, F can be any inclusion monotonic interval extension of f. In particular, equality is allowed in (8.45). In fact, for F defined by equality in (8.45), the assumption (8.41) is satisfied if and only if

$$T[\underline{y}_0, \bar{y}_0](t) \subseteq [\underline{y}_0(t), \bar{y}_0(t)] \quad \text{for all } t \text{ in } [t_0, T].$$

We consider the sequence of interval functions generated by

$$(8.46) \qquad [\underline{y}_{n+1}, \bar{y}_{n+1}] = T[\underline{y}_n, \bar{y}_n].$$

The conclusions expressed in (8.43) now follow from Theorem 5.7. Using equality in (8.45), we find that exactly the same sequences generated by (8.42) are also generated by (8.46). Furthermore, the results expressed in (8.43) follow by use of (8.46) even without equality in (8.45) as long as

$$T[\underline{y}_0, \bar{y}_0](t) \subseteq [\underline{y}_0(t), \bar{y}_0(t)] \quad \text{for all } t \text{ in } [t_0, T].$$

This is true even with $\underline{e} = \bar{e} = 0$ in (8.44) and even without assumptions on f assuring a unique solution.

We now illustrate the remarks just made with a specific example.

Consider the initial value problem

$$(8.47) \qquad y' = ct^2 + y^b, \qquad y(0) = a,$$

where all we know about a, b, and c is that

$$0 \leq a \leq .1, \qquad .2 \leq b \leq .38, \qquad 3.3 \leq c \leq 3.6.$$

Notice that (8.47) does *not* satisfy the usual Lipschitz condition for a unique solution when $a = 0$.

We define the interval operator

$$(8.48) \qquad T(Y)(t) = [0, .1] + \int_0^t \{[3.3, 3.6]s^2 + Y^{[.2, .38]}(s)\} \, ds.$$

Let $Y_0(t) \equiv [0, B]$ for all t in $[0, t_1]$ with $B > 0$ and $t_1 > 0$ to be determined, if possible, such that

$$T(Y_0)(t) \subseteq Y_0(t) \quad \text{for all } t \text{ in } [0, t_1].$$

From (8.48) we find that

$$T[0, B](t) = [0, .1] + \int_0^t \{[3.3, 3.6]s^2 + [0, B]^{[.2,.38]}\}\, ds$$

$$= [0, .1] + [1.1, 1.2]t^3 + [0, B]^{[.2,.38]}t.$$

Now we have

$$[0, B]^{[.2,.38]} = \begin{cases} [0, B^{.2}], & 0 < B \leq 1, \\ [0, B^{.38}], & 1 \leq B. \end{cases}$$

For $B \leq 1$ we will satisfy $T[0, B](t) \subseteq [0, B]$ for all t in $[0, t_1]$ if

(8.49) $.1 + B^{.2}t_1 + 1.2t_1^3 \leq B.$

We can take $B = 1$ in (8.49), for instance, and find that

$$T[0, 1](t) \subseteq [0, 1] \quad \text{for all } t \text{ in } [0, t_1]$$

where t_1 is any positive number such that

(8.50) $t_1 + 1.2t_1^3 \leq .9.$

We can satisfy (8.50) with $t_1 = .6$, for example. It follows, from Theorem 5.7, that every solution to (8.47) is contained in the interval function

$$Y(t) = [0, 0.1] + [0, 1]t + [1.1, 1.2]t^3$$

for all t in $[0, 0.6]$. Put another way, we have

(8.51) $1.1t^3 \leq y(t) \leq .1 + t + 1.2t^3$

for all t in $[0, 0.6]$ for *any* solution to (8.47) with

$$0 \leq a \leq .1, \qquad .2 \leq b \leq .38, \qquad 3.3 \leq c \leq 3.6.$$

The bounds given by (8.51) can be sharpened and continued beyond $t = .6$ by the methods described in §§ 4.5, 5.3, and 8.1.

In addition to the applications of Theorem 5.7, there are other connections between interval mathematics and the theory of differential inequalities as has been pointed out recently by K. Nickel (N 14). An *inverse isotone* operator M has the defining property:

$$Mv \leq Mw \quad \text{implies} \quad v \leq w.$$

For such an operator,

$$Mv \leq r \leq Mw \quad \text{implies} \quad v \leq u \leq w$$

for any solution u of the operator equation

(8.52) $Mu = r.$

Such an equation is said to be of *monotonic type* (N 14).

An example is the following linear parabolic boundary value problem discussed by E. Adams and H. Spreuer (N 15):

(8.53)
$$Mu := u_t - u_{xx} + a(x, t) \, u - f(x, t) = 0$$
$$\text{on } D = \{(x, t): -1 < x < 1, \, 0 < t \le T\};$$
$$u = 0 \quad \text{on } \partial D = \bar{D} \backslash D; \, a, f \in C^{(2+\alpha, 2+\alpha)} (\bar{D});$$
$$f \ge 0 \quad \text{on } \bar{D}; \qquad f(\pm 1, 0) = 0.$$

It can be shown (N 15) that $Mv \le Mw$ on D and $v \le w$ on ∂D imply $v \le w$ on $\bar{D}$. We now describe the method of Adams and Spreuer for the construction of arbitrarily sharp upper and lower bounds for the solution to (8.53). The description is simplified somewhat by using interval notation.

We discretize $\bar{D}$ with a uniform mesh $\{(x_j, t_i)\}$ so that

$$t_{i+1} - t_i = \Delta t = T/N \quad \text{and} \quad x_{j+1} - x_j = \Delta x = 1/(2n) \quad \text{with}$$

$$t_i = i\Delta t \quad \text{and} \quad x_j = j\Delta x \quad \text{for } i = 0, 1, \cdots, N \quad \text{and} \quad j = -n, \cdots, n.$$

We define the sets

$$D_i = \{(x, t): -1 < x < 1, \, t_i < t \le t_{i+1}\}.$$
$$D_{ji} = \{(x, t): x_j \le x \le x_{j+1}, \, t_i \le t \le t_{i+1}\}.$$

We define the intervals

(8.54)
$$A_{j+1,i+1} = a(x_j, t_i) + \Delta x \frac{\partial a}{\partial x}[D_{ji}] + \Delta t \frac{\partial a}{\partial t}[D_{ji}],$$

$$F_{j+1,i+1} = f(x_j, t_i) + \Delta x \frac{\partial f}{\partial x}[D_{ji}] + \Delta t \frac{\partial f}{\partial t}[D_{ji}]$$

where

$$\frac{\partial a}{\partial x}[D_{ji}] = \left[\min_{D_{ji}} \frac{\partial a}{\partial x}, \max_{D_{ji}} \frac{\partial a}{\partial x} \right],$$

etc.

Next, we define the interval functions

$$A_j(t) = \left(1 - \frac{t - t_i}{\Delta t}\right)A_{ji} + \left(\frac{t - t_i}{\Delta t}\right)A_{j,i+1} \quad \text{on } D_i,$$

$$F_j(t) = \left(1 - \frac{t - t_i}{\Delta t}\right)F_{ji} + \left(\frac{t - t_i}{\Delta t}\right)F_{j,i+1} \quad \text{on } D_i,$$

(8.55)
$$A(x, t) = \left(1 - \frac{x - x_j}{\Delta x}\right)A_j(t) + \left(\frac{x - x_j}{\Delta x}\right)A_{j+1}(t) \quad \text{on } D_{ji},$$

$$F(x, t) = \left(1 - \frac{x - x_j}{\Delta x}\right)F_j(t) + \left(\frac{x - x_j}{\Delta x}\right)F_{j+1}(t) \quad \text{on } D_{ji}.$$

We find that, for (x, t) in D_i,

(8.56)
$$w(A(x, t)) \leq \max_{\bar{D}_i \cup \bar{D}_{i-1}} \left\{ \Delta x^2 \left| \frac{\partial^2 a}{\partial x^2} \right| + \Delta t^2 \left| \frac{\partial^2 a}{\partial t^2} \right| \right\},$$

$$w(F(x, t)) \leq \max_{\bar{D}_i \cup \bar{D}_{i-1}} \left\{ \Delta x^2 \left| \frac{\partial^2 f}{\partial x^2} \right| + \Delta t^2 \left| \frac{\partial^2 f}{\partial t^2} \right| \right\}.$$

Now, for y in $S := C^{(2+\alpha,\alpha)}(\bar{D}) \cap C^{(2+\alpha,\infty)}(\bar{D}_i)$, we define the interval operator (mapping real valued functions into interval valued functions)

(8.57) $$Py := y_t - y_{xx} + A(x,t)y - F(x,t).$$

Since $a(x, t) \in A(x, t)$ and $f(x, t) \in F(x, t)$, it follows, using (8.53), that

(8.58) $$My \in Py.$$

Thus, $Py \geq 0$ implies $My \geq 0$ and $Py \leq 0$ implies $My \leq 0$.
    Let $\underline{y}, y, \bar{y} \in S$. If

(8.59) $$P\underline{y} \leq My = 0 \leq P\bar{y} \quad \text{on } D$$

and $\underline{y} = y = \bar{y}$ on ∂D, then $\underline{y} \leq y \leq \bar{y}$ on $\bar{D}$ since then we will have $M\underline{y} \leq My = 0 \leq M\bar{y}$.
    E. Adams and H. Spreuer (N 15) construct finitely representable functions $\underline{y}$ and $\bar{y}$ which are cubic splines in x and which satisfy (8.59). In this way one can compute upper and lower bounds, with second order convergence properties, to the solution of (8.53).
    Numerous other problems "of monotonic type" with inverse isotone operators have been discussed in the literature. H. Spreuer (N 26) presents a method for constructing upper and lower bounds to solutions of linear *elliptic* boundary value problems of the form

(8.60)
$$pu := -\Delta u + c(x)u = f(x)$$
$$\text{in } D = \{x : 0 < x_j < 1, j = 1, 2, \cdots, n\},$$
$$u(x) = 0 \quad \text{for } x \text{ on } \partial D.$$

The problem (8.60) is of monotonic type if $c \geq 0$. Again, the numerical bounds are computed using spline functions.
    A large class of problems of monotonic type has been discussed by J. Schröder (N 16). These include initial value problems, boundary value problems for certain partial differential equations, and integro-differential equations of the form

(8.61) $$-u''(x) + g(x, u(x), u'(x)) + \int_{c(x)}^{d(x)} K(x,t,u(x),u(t),u'(x), u''(t)) \, dt = 0$$

for $0 < x < 1$, where u, g, and K are vector valued functions.

U. Marcowitz (N 17) has used interval methods in combination with the methods of J. Schröder to compute rigorous error bounds for numerical solutions of the re-entry problem for an Apollo type spacecraft. The bounding methods for problems of type (8.61) discussed by Schröder (N 16) include the following applications:

(1) A boundary value problem describing the *bending of a beam*:

$$(8.62) \qquad -u'' - ku(1 - u'^2)^{1/2} = 0, \qquad u(0) = u'(1) = 0;$$

is there a nonnegative solution u for $k > \pi^2/4$?

(2) A *nonlinear eigenvalue problem occurring in the study of vibrations of aircraft structures*:

$$(8.63) \qquad u^{iv} + ((u'')^3)'' - ku = 0 \qquad (0 < x < 1)$$

$$k > \pi^4, \qquad u(0) = u''(0) = u(1) = u''(1) = 0;$$

is there a nonnegative solution u?

(3) The *Fermi–Thomas problem, occurring in atomic physics*:

$$(8.64) \qquad -u'' + x^{-1/2} u^{3/2} = 0 \qquad (0 < x < \infty),$$

$$u(0) = 1, \qquad \lim_{x \to \infty} u(x) = 0.$$

The application of interval methods to the *computation* of the *defect* in an approximate solution (for instance $P(y)$ in (8.57) for an approximate solution y for (8.53)) to a boundary value problem in *partial differential equations* was first pointed out by F. Krückeberg (N 18).

An interesting discussion of *applications* of methods based on the theory of differential inequalities and also those based on interval analysis to *dynamical models of ecosystems* with uncertain parameters has recently been given by G. W. Harrison (N 20); he has pointed out correctly the importance of dealing with a difficulty, which has come to be known (N 21) as the "wrapping effect", inherent in all methods for following *sets of solutions* to initial value problems with uncertain parameters or uncertain initial values (see § 8.1).

Using concepts and notation from interval mathematics, K. Nickel (N 23) has recently clarified a "lemma" of M. Muller and extended it to functional-differential equations of the type

$$(8.65) \qquad \begin{aligned} u'(t) &= f(t, u(t), u), \qquad 0 < t \leq T, \\ u(0) &= a \end{aligned}$$

where u, f, and a are n-dimensional vectors and, for each k and t, f_k $(t, u(t), u)$ is a "Volterra" functional on u depending only on the values $u(s)$ for $0 \leq s \leq t$.

Using the notation

$$(8.66) \qquad {}_k u = (u_1, u_2, \cdots, u_{k-1}, u_{k+1}, \cdots, u_n),$$

Nickel proves the following.

Suppose that v and w are differentiable functions on $(0, T]$ such that $v \leqq w$ and $v(0) < a < w(0)$ and, for all t in $(0, T]$ and each k, we have

$$v'_k(t) < f_k(t, y^{(k)}(t), [v, w]) \quad \text{for all } y^{(k)} \text{ such that}$$

$$_k v \leqq {_k}y^{(k)} \leqq {_k}w \quad \text{and} \quad y_k^{(k)} = v_k \quad \text{and}$$

(8.67)

$$w'_k(t) > f_k(t, z^{(k)}(t), [v, w]) \quad \text{for all } z^{(k)} \text{ such that}$$

$$_k v \leqq {_k}z^{(k)} \leqq {_k}w \quad \text{and} \quad z_k^{(k)} = w_k.$$

Then any solution u of (8.65) is bounded by

(8.68) $$v(t) \leqq u(t) \leqq w(t) \quad \text{for all } t \text{ in } [0, T].$$

This result, which resembles an application of Theorem 5.7 to (8.65), but does not seem to follow from Theorem 5.7, has also been extended by Nickel to *strongly coupled systems of parabolic function-differential equations* (N 23).

E. Adams and W. F. Ames (N 31) obtain contracting interval iteration methods for nonlinear problems in R^n by constructing an auxiliary, inverse isotone operator in R^{2n}. The methods are then applied to discretizations of boundary value problems.

NOTES

1. See W. F. Ames and E. Adams [70, pp. 135–140].

2. See, for example, R. E. Moore [56], [57], [58], [59], [70, pp. 31–47], [76, Vol I, pp. 61–130; Vol. II, pp. 103–140]; F. Krückeberg [25, pp. 91–97]; K.-H. Bachmann [8]; J. W. Daniel and R. E. Moore [11]; J. Avenhaus [7]; S. Hunger [34]; N. F. Stewart [91], [93].

A very interesting discussion of an initial value problem—actually a *control theory problem* concerning the re–entry of an Apollo-type spaceship—has been given by U. Marcowitz [53]. Interval computation was used, along with methods of functional analysis, in a number of places to help obtain a numerical solution with error bounds using a "defect" procedure. Here, as elsewhere, interval computation provides a convenient means of obtaining bounds on the ranges of values of functions, bounds on norms of operators, and bounds on roundoff error.

3. U. Wauschkuhn [98].

4. R. E. Moore [25, pp. 67–73].

5. R. E. Moore [70, pp. 41–42].

6. E. Hansen [25, pp. 74–90].

7. R. E. Moore [62], [70, pp. 31–47] and also F. A. Oliveira [72].

8. W. Neuland [65].

9. W. Appelt [6].

10. F. Krückeberg [25, pp. 98–101] and also R. Tost [96].

11. See R. E. Moore, *Two-sided approximation to solutions of nonlinear operator equations—a comparison of methods from classical analysis, functional analysis, and interval analysis* [70, pp. 31–47].

12. See W. F. Ames and E. Adams, *Monotonically convergent numerical two-sided bounds for a differential birth and death process* [70, pp. 135–140].

13. See G. Scheu and E. Adams, *Zur numerischen Konstruktion konvergenter Schrankenfolgen für Systeme nichtlinear, gewöhnlicher Anfangswertaufgaben* [70, pp. 279-287].

14. See K. Nickel [65a].

15. See E. Adams and H. Spreuer, *Konvergente numerische Schrankenkonstruktion mit Spline-Funktionen für nichtlineare gewöhnliche bzw. lineare parabolische Randwertaufgaben* [70, pp. 118-126].

16. See J. Schröder [86a], [86b].

17. See U. Markowitz [53].

18. See F. Krückeberg [46] and [25, pp. 98-101].

19. See W. F. Ames and M. Ginsberg [3a].

20. G. W. Harrison (Mathematics Department and Institute of Ecology, University of Georgia, Athens, GA) *Dynamic models with uncertain parameters*, presented at the First International Conference on Mathematical Modeling, St. Louis, 1977. See also [30a].

21. See, N. F. Stewart [91].

22. See R. E. Moore [57] and [76, Vol. I, pp. 61-130].

23. See K. Nickel [65b], [65c], R. Lohner and E. Adams [48a].

24. See D. P. Davey and N. F. Stewart [12]; F. Krückeberg [25, pp. 91-97]; R. E. Moore [76, Vol. II, pp. 103-140]; N. F. Stewart [91]; R. Lohner and E. Adams [48a]; L. W. Jackson [36].

25. See J. W. Daniel and R. E. Moore [11].

26. See H. Spreuer, *Konvergente numerische Schranken für partielle Randwertaufgaben von monotoner Art*, [70, pp. 298-305].

27. See G. W. Harrison [30a].

28. See D. Greenspan [21a].

29. See R. E. Moore [59, Appendix C].

30. Y. F. Chang, G. Corliss and G. Kriegsman, Dept. of Computer Science, Dept. of Math. and Statistics, University of Nebraska, personal communication.

31. See E. Adams and W. F. Ames [1].

Chapter **9**

An Application in Finance: Internal Rates-of-Return

One way of measuring the profitability of a proposed investment is to compute the project's *internal rate-of-return* (IRR). The IRR, r^*, is a solution of the equation

$$C_0 + C_1(1+r)^{-1} + \cdots + C_N(1+r)^{-N} = R_0 + R_1(1+r)^{-1} + \cdots + R_N(1+r)^{-N}$$

where C_i is the investment made in the project in year i and R_i is the return received from the project in year i.

We can define the *net transfer* $n_i = C_i - R_i$ in the ith year (investments minus returns). It will be assumed that all transactions are made on the first of the year. Actually, any time period could be used in place of a year (a month, a day, etc.). The present value of the total project at discount rate r at the start of the year i is $p_i(r)/(1+r)^i$ where the polynomials $p_i(r)$ satisfy the recursion relation

(9.1) $$p_i(r) = n_i + (1+r)p_{i-1}(r) \quad \text{for } i = 1, 2, \cdots, N,$$

where $p_0(r) = n_0 > 0$ for a positive initial investment.

We can also express the sought-after internal rate-of-return r^* as a root of the polynomial

(9.2) $$p_N(r) = n_N + n_{N-1}(1+r) + \cdots + n_0(1+r)^N.$$

We assume that n_N is negative so that there is a final return to close out the investment. The net transfers $n_1, n_2, \cdots, n_{N-1}$ can be of either sign (positive or negative or even zero) with as many sign changes as desired.

The same polynomials $p_i(r)$ and $p_N(r)$ defined by (9.1) and (9.2) can also be interpreted as the current balances at years i (or N, respectively) for a savings account. In this analogy, the net transfer n_i would be the difference between deposits and withdrawals at the beginning of year i. Here, r would be the (annual) interest paid on the account, to be closed out at year N. The ordinary requirement that the current balance in a savings account be kept nonnegative

117

can be expressed by the conditions

(9.3) $p_i(r^*) \geqq 0$ for all $i = 0, 1, 2, \cdots, N$,

where r^* is the actual interest rate paid. The analogous condition for a proposed investment would be

$$(C_0 - R_0) + (C_1 - R_1)(1 + r^*)^{-1} + \cdots + (C_k - R_k)(1 + r^*)^{-k} \geqq 0$$

for all $k = 0, 1, 2, \cdots, N$.

We will *assume* the conditions (9.3) in this chapter. If we *find* a root of (9.2) we can *check* the conditions (9.3). Under the assumptions that $n_0 > 0$ and $n_N < 0$, the polynomial $p_N(r)$ given by (9.2) will clearly have at least one real root in $(-1, \infty)$ since $p_N(-1) = n_N < 0$ and $p_N(r) > 0$ for very large positive r. Note that r^* could be positive or negative (but > -1).

We now prove the following useful result.

LEMMA 9.1. *The polynomial* (9.2) *has at most one root* r^* *in* $(-1, \infty)$ *which satisfies* (9.3). *If it has such a root* r^*, *then there are* no *roots* (*whether or not they satisfy* (9.3)) *beyond it* ($> r^*$) *and Newton's method converges to* r^* *from any* $r_0 \geqq r^*$.

Proof. Suppose that $p_N(r^*) = 0$ for some $r^* > -1$, and that $p_N(r^*)$ satisfies (9.3). By repeated differentiation of (9.1) we find that

$$p_0'(r) = 0,$$

(9.4)
$$p_i'(r) = p_{i-1}(r) + (1 + r)p_{i-1}'(r), \qquad i = 1, \cdots, N;$$

$$p_i^{(k)}(r) = 0 \quad \text{for } i = 0, 1, \cdots, k - 1,$$

$$p_i^{(k)}(r) = kp_{i-1}^{(k-1)}(r) + (1 + r)p_{i-1}^{(k)}(r), \qquad i = k, \cdots, N.$$

It follows from (9.3) that, for $r = r^*$, we have (since $p_0(r^*) > 0$)

(9.5) $p_N^{(k)}(r^*) > 0$ for $k = 1, 2, \cdots, N$.

Thus, $p_N(r)$ can be represented in the form (since $p_N(r^*) = 0$)

(9.6)
$$p_N(r) = \sum_{j=1}^{N} (1/k!)p_N^{(k)}(r^*)(r - r^*)^k$$

and so $p_N(r) > 0$ for $r > r^*$. Similarly, we have

(9.7) $p_N'(r) > 0$ for $r > r^*$ and $p_N''(r) > 0$ for $r > r^*$.

It follows from $p_N'(r) > 0, p_N''(r) > 0, p_N(r) > 0$ for $r > r^*$ that there is no real root of $p_N(r)$ greater than r^* and that Newton's method converges to r^* from any $r_0 > r^*$. There cannot be two distinct values of r^* which are roots of $p_N(r)$ and satisfy (9.3) for then we would have a contradiction of (9.7) for some r between them. This completes the proof of the lemma.

Before we discuss the application of Lemma 9.1 to the analysis of internal rates-of-return, it is worth noting that by the evaluation of (9.1) in interval arithmetic for some trial interval, R, can show that there is *no root* in R which satisfies (9.3) if, for some i, we have $p_i(R) < 0$. We can use such a test to search for intervals R which *may* contain a root of (9.2) which satisfies (9.3) by excluding parts of the interval $(-1, \infty)$ which do not (N 1). Note further that $p_N(0) > 0$ implies that there is a real root of $p_N(r)$ in $(-1, 0)$; whereas $p_N(0) = n_N + n_{N-1} + \cdots + n_0 < 0$ implies that there is a positive real root of $p_N(r)$. Put another way, if the sum of investments (costs) is less than the sum of returns (benefits) then there is a positive rate-of-return. It is not hard to show that if there is a positive rate-of-return r^* which satisfies (9.3), then $p_N(0) < 0$. Therefore, *if $n_N + n_{N-1} + \cdots + n_0 \geq 0$, then there is no positive rate-of-return r^* which satisfies (9.3).* We can show this as follows. If $p_N(r^*) = 0$ for $r^* > 0$, with $p_i(r^*) \geq 0$ for $i = 1, 2, \cdots, N-1$ and $p_0(r) = n_0 > 0$, then it follows that (from (9.1)):

$$n_N + (1 + r^*)p_{N-1}(r^*) = 0$$

so

$$-n_N/p_{N-1}(r^*) = 1 + r^* > 1 \quad \text{and} \quad 0 \leq p_{N-1}(r^*) < -n_N.$$

Similarly, $0 \leq n_{N-1} + (1 + r^*)p_{N-2}(r^*) < -n_N$ implies

$$0 < r^* < (-n_N - n_{N-1} - p_{N-2}(r^*))/p_{N-2}(r^*);$$

therefore

$$0 \leq p_{N-2}(r^*) < -n_N - n_{N-1}.$$

Continuing in this way, we obtain $0 < n_0 < -n_N - n_{N-1} - \cdots - n_1$. This shows that $p_N(0) = n_N + n_{N-1} + \cdots + n_0 < 0$ is a *necessary condition* for the existence of a positive rate-of-return r^* which satisfies the conditions (9.3).

We next discuss the application of Lemma 9.1 to a *real world problem* arising in connection with the World Bank's evaluation of a proposed investment in a forestry project (N 1). The projected cost-benefit stream was as follows:

$$n_0 = 40{,}000.$$

$$n_1 = 20{,}000.$$

$$n_2 = 10{,}000.$$

$$n_3 = \cdots = n_{18} = 5{,}000.$$

$$n_{19} = -480{,}000.$$

$$n_{20} = 20{,}000.$$

$$n_{21} = 10{,}000.$$

$$n_{22} = \cdots = n_{38} = 5,000.$$

$$n_{39} = -480,000.$$

$$n_{40} = 10,000.$$

$$n_{41} = \cdots = n_{48} = 5,000.$$

$$n_{49} = -500,000.$$

Notice that, here, we have $N = 49$ and that there are five changes of sign in the cost-benefit stream. After computing that $p_{49}(0) < 0$, we know that there is a positive root of $p_{49}(r)$. We try $r_0 = 0.1$, 0.2, etc. until $p_{49}(r_0) > 0$. We find that $p_{49}(0.1) > 0$ and so we try $r_0 = 0.1$ as a starting point in Newton's method for finding a root of $p_{49}(r)$ with the coefficients as given. We use the recursive evaluation of polynomial values and derivative values as given by (9.1) and (9.4) and program Newton's method, for this example in FORTRAN as follows: (with P for p, PP for p', and C(I) for n_I)

```
        DIMENSION  C(50)
        DATA C/40., 20., 10., 16*5., -480., 20., 10., 17*5.,
    $              -480., 10., 8*5., -500./
        K = 0
        R = .1
   10   P = C(1)
        PP = 0.
        DO 20 I = 2, 50
        PP = P + (1. + R)*PP
   20   P = C(I) + (1. + R)*P
        PRINT*, K, R, P, PP
        R1 = R - P/PP
        IF(R1. GE. R) STOP
        R = R1
        K = K + 1
        GO TO 10
        END
```

(Note that the *scaling* of the coefficients does not appreciably affect the results.)

The execution of this program converged monotonically downward in *four iterations* (on the University of Wisconsin UNIVAC 1110 computer) to the value ($R = r^*$, given here to five places) $r^* = .09116$.

We can check that $p_i(r^*) \geqq 0$ as in (9.3) for $i = 0, 1, \cdots, 49$. We will return to this point later. The stopping criterion $r_{k+1} \geqq r_k$ for a monotonically decreasing (mathematically) sequence $\{r_k\}$, computed in finite precision machine arithmetic, will always be successful because there are only a finite number of machine numbers (in any finite precision representation). Put another way, roundoff error will cause the machine-computed sequence to lose monotonicity after some *finite number* of steps—*four* in the example at hand.

Another question of interest in connection with rates-of-return is that of the *sensitivity* of the rate-of-return with respect to percentage variation (plus or minus) in the coefficients of the cost-benefit stream. If the values of n_i for $i = 0, 1, \cdots, N$ are allowed to vary independently plus or minus 5%, 10%, or 15%, then how much will the rate-of-return vary?

In this case, we can regard the *coefficients* of the polynomial $p_N(r)$ given by (9.2) or by (9.1) as *intervals*: $n_i = [\underline{n}_i, \bar{n}_i]$, with

$$\underline{n}_i = n_i^* - a|n_i^*| \quad \text{and} \quad \bar{n}_i = n_i^* + a|n_i^*|$$

for $i = 0, 1, \cdots, N$. For 5%, 10%, or 15% variations in the coefficients, we would put $a = .05, .10$, or $.15$, respectively. Here, n_i^* for $i = 0, 1, \cdots, N$ are the coefficients as originally given (leading to a rate of return r^*).

The resulting *interval polynomial* $P(r)$ can be represented, for values of $r > -1$ (so that $(1+r)$ is positive) by

(9.8) $$P(r) = [\underline{P}(r), \bar{P}(r)]$$

where

$$\underline{P}(r) = \sum_{i=0}^{N} \underline{n}_i(1+r)^i \quad \text{and} \quad \bar{P}(r) = \sum_{i=0}^{N} \bar{n}_i(1+r)^i.$$

We can compute the values of the polynomials $\underline{P}(r)$ and $\bar{P}(r)$, and their derivatives $\underline{P}'(r)$ and $\bar{P}'(r)$ using

(9.9)
$$\underline{P}_i(r) = \underline{n}_i + (1+r)\underline{P}_{i-1}(r), \qquad \underline{P}_0(r) = \underline{n}_0;$$
$$\bar{P}_i(r) = \bar{n}_i + (1+r)\bar{P}_{i-1}(r), \qquad \bar{P}_0(r) = \bar{n}_0;$$
$$\underline{P}'_i(r) = \underline{P}_{i-1}(r) + (1+r)\underline{P}'_{i-1}(r), \qquad \underline{P}'_0(r) = 0;$$
$$\bar{P}'_i(r) = \bar{P}_{i-1}(r) + (1+r)\bar{P}'_{i-1}(r), \qquad \bar{P}'_0(r) = 0;$$

with $\underline{P}(r) = \underline{P}_N(r)$ and $\bar{P}(r) = \bar{P}_N(r)$.

A *fast way* to find the range of values $[\underline{r}, \bar{r}]$ of the rate-of-return corresponding to variations in the coefficients of the cost-benefit stream, as represented above, is the following. First we find r^* such that $p_N(r^*) = 0$ (approximately) using $p_N(r)$, with coefficients n_i^*, $i = 0, 1, \cdots, N$, given by (9.2) or (9.1). Next, we use the iteration formulas

(9.10)
$$\underline{r}_{k+1} = \underline{r}_k - \bar{P}(\underline{r}_k)/\bar{P}'(\underline{r}_k), \qquad \underline{r}_0 = r^*;$$
$$\bar{r}_{k+1} = \bar{r}_k - \underline{P}(\bar{r}_k)/\underline{P}'(\bar{r}_k), \qquad \bar{r}_0 = r^*.$$

We iterate the formulas (9.10) until $k \geq 2$ *and* $\bar{r} = \bar{r}_{k+1} \geq \bar{r}_k$. We then put $\underline{r} = \underline{r}_{k+1}$ as well. If the variations of the coefficients in the cost-benefit stream are reasonably small, then this procedure should give rapid convergence to the range of values of possible rates-of-return. If we carry this procedure out for several different percentage variations, we will get a picture of the *sensitivity* of the rate-of-return with respect to variations in the cost-benefit stream.

Again, we can check that the conditions (9.3) are satisfied for the resulting roots $\underline{r}$ and $\bar{r}$ (of the polynomials $\bar{P}(r)$ and $\underline{P}(r)$, respectively). It is not necessary that (9.3) be satisfied for the *iterates* during the computation of $\underline{r}$ and $\bar{r}$, but only for the final "converged" values. In this case we will have good approximations to the actual range of values of the rate-of-return.

The procedure described above has been applied to the forestry project, discussed earlier in this chapter, with the following results.

% variation in n_i	$\underline{r}$	$\bar{r}$	# iterations of (9.10)
5%	.085016	.09737	4
10%	.078891	.10366	7
15%	.072760	.11007	6

The conditions (9.3) *are* satisfied for these results, produced from the following FORTRAN program.

```
     DIMENSION C(50), CL(50), CU(50)
     DATA C/40., 20., 10., 16*5., −480., 20., 10., 17*5.,
   $            −480., 10., 8*5., −500./
     DO 30 J = 1, 3
     A = J*.05
     DO 5 I = 1, 50
     CL(I) = C(I) − A*ABS(C(I))
   5 CU(I) = C(I) + A*ABS(C(I))
     K = 0
     RL = .09116
     RU = .09116
  10 PL = CL(1)
     PU = CU(1)
     PPL = 0.
     PPU = 0.
     DO 20 I = 2, 50
     PPL = PL + (1. + RU)*PPL
     PPU = PU + (1. + RL)*PPU
     PL = CL(I) + (1. + RU)*PL
     PU = CU(I) + (1. + RL)*PU
  20 PRINT*, PL, PU
     PRINT*, K, RL, RU
     R1 = RL − PU/PPU
     R2 = RU − PL/PPL
```

```
      IF(K.GE.2.AND.R2.GE.RU)GO TO 30
      RL = R1
      RU = R2
      K = K + 1
      GO TO 10
30    CONTINUE
      END
```

In this program, we use the notation CL(I) for $\underline{n}_i$, CU(I) for $\bar{n}_i$, RL for $\underline{r}$, RU for $\bar{r}$, PL for $\underline{P}$, PU for $\bar{P}$, PPL for $\underline{P}'$, and PPU for $\bar{P}'$.

It should be clear that the polynomial $\underline{P}(r)$ lies *below* the polynomial $p_N(r)$ and will cross the r axis to the *right* of r^*; while the polynomial $\bar{P}(r)$ lies *above* the polynomial $p_N(r)$ and will cross the r-axis to the *left* of r^* This is why RU is used in the iteration formulas with PL and PPL whereas RL is used in the iteration formulas with PU and PPU.

NOTES

1. P. Gutterman, Computer Activities Dept., The World Bank, Washington, DC, personal communication.

Appendix: A BIBLIOGRAPHY ON INTERVAL–MATHEMATICS,*

by Fritz Bierbaum† and Klaus-Peter Schwiertz††

At the appearance of this bibliography we want to thank warmly all those who helped us in collecting the titles by sending us reprints, preprints, and manuscripts of unpublished papers. The bibliography on interval–mathematics contains about 760 titles arranged alphabetically by the names of the authors. The publications of the same author are arranged by the year of publication. The abbreviations of the journals are in agreement with those prescribed in Mathematical Reviews 44, 1606-1628 (1972). If the name of a journal has changed the old name is inserted next to the new in parentheses, (for example: Angewandte Informatik (Elektron. Datenverarbeitung)). Citations from books have been designed in the following manner: title of the chapter, 'title of the book', editor, publisher, number of pages (year). The abbreviation GMD denotes the Gesellschaft für Mathematik und Datenverarbeitung, D-5202 Birlinghoven bei Bonn, Schloss.

At the Institut für Angewandte Mathematik der Universität Freiburg there exists an "Internal-Library" containing nearly all the titles of the bibliography. It is open to all users.

This library and the bibliography are attended now by the second author. It is always possible to get a continuation of the bibliography from him.

We want to thank all readers sending us corrections of the bibliography. Finally, we would like to ask all those working in interval-mathematics to send any new papers to the second author.

---

*Reprinted in its entirety from *Journal of Computational and Applied Mathematics,* 4 (1978), pp. 59-86.

†Institut für Praktische Mathematik der Universität Karlsruhe, Englerstr. 2, D-7500 Karlsruhe, Fed. Rep. of Germany.

††Institut für Angewandte Mathematik der Universität Freiburg i.Br., Hermann-Herderstr. 10, D-7800 Freiburg i.Br., Fed. Rep. of Germany.

ABERTH-ALEFELD

1* ABERTH, O.
A PRECISE NUMERICAL ANALYSIS PROGRAM
COMM. ACM 17, 509-513 (1974)

ADAMS, E.
SIEHE AUCH: AMES, W. F.; ADAMS, E.
FAASS, E.; ADAMS, E.
SCHEU, G.; ADAMS, E.
SPREUER, H.; ADAMS, E.
SPREUER, H.; ADAMS, E.;
SRIVASTAVA, V. N.

2* ADAMS, E.; SCHEU, G.
ZUR NUMERISCHEN DURCHFUEHRUNG EINES
ITERATIONSVERFAHRENS FUER MONOTONE SCHRANKENFOLGEN
BEI NICHTLINEAREN GEWOEHNLICHEN RAND- ODER
ANFANGSWERTAUFGABEN
Z. ANGEW. MATH. MECH. 56, T270-T272 (1976)

3* ADAMS, E.; SPREUER, H.
UEBER SCHRANKEN DER LOESUNGEN DER
GRENZSCHICHT-GLEICHUNGEN FUER
HYPERSCHALLSTROEMUNGEN
Z. ANGEW. MATH. MECH. 51, T141-T143 (1971)

4* ADAMS, E.; SPREUER, H.
KONVERGENTE NUMERISCHE SCHRANKENKONSTRUKTION MIT
SPLINE-FUNKTIONEN FUER NICHTLINEARE GEWOEHNLICHE
BZW. LINEARE PARABOLISCHE RANDWERTAUFGABEN
'INTERVAL MATHEMATICS', ED. BY K. NICKEL
LECTURE NOTES IN COMPUTER SCIENCE 29, SPRINGER
VERLAG, 118-126 (1975)

5* ADAMS, E.; SPREUER, H.
UEBER DAS VORLIEGEN DER EIGENSCHAFT VON MONOTONER
ART BEI FORTSCHREITEND BZW. NUR ALS GANZES
LOESBAREN SYSTEMEN
Z. ANGEW. MATH. MECH. 55, T191-T193 (1975)

6* AHMAD, R.
A DISTRIBUTION-FREE INTERVAL MATHEMATICAL ANALYSIS
OF PROBABILITY DENSITY FUNCTIONS
'INTERVAL MATHEMATICS', ED. BY K. NICKEL
LECTURE NOTES IN COMPUTER SCIENCE 29, SPRINGER
VERLAG, 127-134 (1975)

7* ALBRECHT, J.
MONOTONE ITERATIONSFOLGEN UND IHRE VERWENDUNG ZUR
LOESUNG LINEARER GLEICHUNGSSYSTEME
NUMER. MATH. 3, 345-358 (1961)

8* ALBRECHT, J.
FEHLERSCHRANKEN UND KONVERGENZBESCHLEUNIGUNG BEI
EINER MONOTONEN ODER ALTERNIERENDEN
ITERATIONSFOLGE
NUMERISCHE MATHEMATIK 4, 196-208 (1962)

9* ALBRECHT, J.
INTERVALLSCHACHTELUNGEN BEIM ROMBERGVERFAHREN
Z. ANGEW. MATH. MECH. 52, 433-435 (1972)

10* ALBRECHT, J.
INTERVALLSCHACHTELUNGEN BEIM ROMBERGVERFAHREN
(NOTIZ)
Z. ANGEW. MATH. MECH. 53, T176 (1973)

11* ALBRECHT, R.
GRUNDLAGEN EINER THEORIE GERUNDETER ALGEBRAISCHER
VERKNUEPFUNGEN IN TOPOLOGISCHEN VEREINEN
COMPUTING SUPPLEMENTUM 1, ALBRECHT, R.;
KULISCH, U., SPRINGER-VERLAG WIEN, NEW YORK,
1-14 (1977)

12* ALBRECHT, R.; KULISCH, U.
GRUNDLAGEN DER COMPUTER-ARITHMETIK
COMPUTING SUPPLEMENTUM 1, SPRINGER-VERLAG, WIEN;
NEW YORK (1977)

13* ALEFELD, G.
INTERVALLRECHNUNG UEBER DEN KOMPLEXEN ZAHLEN UND
EINIGE ANWENDUNGEN
DISSERTATION, UNIVERSITAET KARLSRUHE (1968)

14* ALEFELD, G.
UEBER DIE AUS MONOTON ZERLEGBAREN, OPERATOREN
GEBILDETEN ITERATIONSVERFAHREN
COMPUTING 6, 161-172 (1970)

15* ALEFELD, G.
UEBER EIGENSCHAFTEN UND ANWENDUNGSMOEGLICHKEITEN
EINER KOMPLEXEN INTERVALLARITHMETIK
Z. ANGEW. MATH. MECH. 50, 455-465 (1970)

16* ALEFELD, G.
EINE MODIFIKATION DES NEWTONVERFAHRENS ZUR
BESTIMMUNG DER REELLEN NULLSTELLEN EINER REELLEN
FUNKTION
Z. ANGEW. MATH. MECH. 50, T32-T33 (1970)

17* ALEFELD, G.
BEMERKUNGEN ZUR EINSCHLIESSUNG DER LOESUNG EINES
LINEAREN GLEICHUNGSSYSTEMS
Z. ANGEW. MATH. MECH. 50, T33-T35 (1970)

18* ALEFELD, G.
VERALLGEMEINERUNG DES SCHULZSCHEN
ITERATIONSVERFAHRENS
Z. ANGEW. MATH. MECH. 51, T33 (1971)

19* ALEFELD, G.
ANWENDUNGEN DES FIXPUNKTSATZES FUER
PSEUDOMETRISCHE RAEUME IN DER INTERVALLRECHNUNG
NUMER. MATH. 17, 33-39 (1971)

20* ALEFELD, G.
UEBER DIE EXISTENZ EINER EINDEUTIGEN LOESUNG BEI
EINER KLASSE NICHTLINEARER GLEICHUNGSSYSTEME UND
DEREN BERECHNUNG MIT ITERATIONSVERFAHREN
APL. MAT. 17, 267-294 (1972)

21* ALEFELD, G.
ZUR KONVERGENZ DES VERFAHRENS DER TANGIERENDEN
HYPERBELN UND DES TSCHEBYSCHEFF-VERFAHRENS BEI
KONVEXEN ABBILDUNGEN
COMPUTING 11, 379-390 (1973)

22* ALEFELD, G.
STETS KONVERGENTE VERFAHREN HOEHERER ORDNUNG ZUR
BERECHNUNG VON REELLEN NULLSTELLEN
COMPUTING 13, 55-65 (1974)

23* ALEFELD, G.
QUADRATISCH KONVERGENTE EINSCHLIESSUNG VON
LOESUNGEN NICHTKONVEXER GLEICHUNGSSYSTEME
Z. ANGEW. MATH. MECH. 54, 335-342 (1974)

24* ALEFELD, G.
ZUR KONVERGENZ DES PEACEMAN-RACHFORD-VERFAHRENS
NUMER. MATH. 26, 409-419 (1976)

25* ALEFELD, G.
UEBER DIE DURCHFUEHRBARKEIT DES GAUSSSCHEN
ALGORITHMUS BEI GLEICHUNGEN MIT INTERVALLEN ALS
KOEFFIZIENTEN
COMPUTING SUPPLEMENTUM 1, ALBRECHT, R.;
KULISCH, U.; SPRINGER-VERLAG WIEN, NEW YORK,
15-19 (1977)

26* ALEFELD, G.; APOSTOLATOS, N.
AUFLOESUNG NICHTLINEARER GLEICHUNGSSYSTEME MIT
ZWEI UNBEKANNTEN
BER. D. RECHENZENTRUMS U. D. INST. F. ANGEW. MATH.
UNIVERSITAET KARLSRUHE (1967)

27* ALEFELD, G.; APOSTOLATOS, N.
PRAKTISCHE ANWENDUNG VON ABSCHAETZUNSFORMELN BEI
ITERATIONSVERFAHREN
BER. D. RECHENZENTRUMS U. D. INST. F. ANGEW. MATH.
UNIVERSITAET KARLSRUHE (1968)

28* ALEFELD, G.; APOSTOLATOS, N.
PRAKTISCHE ANWENDUNG VON ABSCHAETZUNGSFORMELN BEI
ITERATIONSVERFAHREN
Z. ANGEW. MATH. MECH. 48, T46-T49 (1968)

29* ALEFELD, G.; HERZBERGER, J.
ALGOL-60 ALGORITHMEN ZUR AUFLOESUNG LINEARER
GLEICHUNGSSYSTEME MIT FEHLERERFASSUNG
COMPUTING 6, 28-34 (1970)

30* ALEFELD, G.; HERZBERGER, J.
UEBER DIE BERECHNUNG DER INVERSEN MATRIX MIT HILFE
DER INTERVALLRECHNUNG
ELEKTRON. RECHENANLAGEN 12, 259-261 (1970)

ALEFELD-ALLENDOERFER

31* ALEFELD, G.; HERZBERGER, J.
MATRIZENINVERTIERUNG MIT FEHLERERFASSUNG
ANGEW. INFORMATIK (ELEKTRON. DATENVERARBEITUNG)
12, 410-416 (1970)

32* ALEFELD, G.; HERZBERGER, J.
UEBER DAS NEWTON-VERFAHREN BEI NICHTLINEAREN
GLEICHUNGSSYSTEMEN
Z. ANGEW. MATH. MECH. 50, 773-774 (1970)

33* ALEFELD, G.; HERZBERGER, J.
VERFAHREN HOEHERER ORDNUNG ZUR ITERATIVEN
EINSCHLIESSUNG DER INVERSEN MATRIX
Z. ANGEW. MATH. PHYS. 21, 819-824 (1970)

34* ALEFELD, G.; HERZBERGER, J.
UEBER DIE VERBESSERUNG VON SCHRANKEN FUER DIE
LOESUNG BEI LINEAREN GLEICHUNGSSYSTEMEN
ANGEW. INFORMATIK (ELEKTRON. DATENVERARBEITUNG)
13, 107-112 (1971)

35* ALEFELD, G.; HERZBERGER, J.
ZUR INVERTIERUNG LINEARER UND BESCHRAENKTER
OPERATOREN
MATH. Z. 120, 309-317 (1971)

36* ALEFELD, G.; HERZBERGER, J.
EINSCHLIESSUNGSVERFAHREN ZUR BERECHNUNG DES
INVERSEN OPERATORS
Z. ANGEW. MATH. MECH. 52, T197-T198 (1972)

37* ALEFELD, G.; HERZBERGER, J.
NULLSTELLENEINSCHLIESSUNG MIT DEM NEWTON-VERFAHREN
OHNE INVERTIERUNG VON INTERVALLMATRIZEN
NUMER. MATH. 19, 56-64 (1972)

38* ALEFELD, G.; HERZBERGER, J.
EIN VERFAHREN ZUR MONOTONEN EINSCHLIESSUNG VON
LOESUNGEN NICHTLINEAR GLEICHUNGSSYSTEME
Z. ANGEW. MATH. MECH. 53, T176-T177 (1973)

39* ALEFELD, G.; HERZBERGER, J.
UEBER DIE KONVERGENZORDNUNG
INTERVALLARITHMETISCHER ITERATIONSVERFAHREN
KOLLOQUIUM UEBER KONSTRUKTIVE THEORIE DER
FUNKTIONEN, TAGUNGSBERICHT, CLUJ (1973)
PUBLIZIERT IN: REVUE D'ANALYSE NUMERIQUE ET DE LA
THEORIE DE L'APPROXIMATION (CLUJ) 3, 117-124
(1974)

40* ALEFELD, G.; HERZBERGER, J.
UEBER DIE VERBESSERUNG VON SCHRANKEN FUER DIE
EIGENWERTE SYMMETRISCHER TRIDIAGONALMATRIZEN
ANGEW. INFORMATIK (ELEKTRON. DATENVERARBEITUNG)
16, 27-35 (1974)

41* ALEFELD, G.; HERZBERGER, J.
EINFUEHRUNG IN DIE INTERVALLRECHNUNG
HERAUSG. K.H. BOEHLING, U. KULISCH, H. MAURER,
BIBLIOGRAPHISCHES INSTITUT MANNHEIM/WIEN/ZUERICH
(1974)

42* ALEFELD, G.; HERZBERGER, J.
UEBER SIMULTANVERFAHREN ZUR BESTIMMUNG REELLER
POLYNOMWURZELN
Z. ANGEW. MATH. MECH. 54, 413-420 (1974)

43* ALEFELD, G.; HERZBERGER, J.
ON THE CONVERGENCE SPEED OF SOME ALGORITHMS FOR
THE SIMULTANEOUS APPROXIMATION OF POLYNOMIAL ROOTS
SIAM J. NUMER. ANAL. 11, 237-243 (1974)

44* ALEFELD, G.; HERZBERGER, J.; MAYER, O.
UEBER NEUERE GESICHTSPUNKTE BEIM NUMERISCHEN
RECHNEN
MATH. NATURW. UNTERRICHT 24, 458-467 (1971)

45* ALLENDOERFER, U.
ABBRECHKRITERIUM UND NUMERISCHE KONVERGENZ IN
HALBGEORDNETEN MENGEN
DIPLOMARBEIT AM INST. F. PRAKT. MATH. DER UNIV.
KARLSRUHE (1976)

AMANN-APPELT

49* AMES, W. F.; GINSBERG, M.

46* AMANN, H.
FIXED POINT EQUATIONS AND NONLINEAR EIGENVALUE
PROBLEMS IN ORDERED BANACH SPACES
SIAM REVIEW 18, 620-709 (1976)

47* AMES, W. F.; ADAMS, E.
MONOTONICALLY CONVERGENT NUMERICAL TWO-SIDED
BOUNDS FOR A DIFFERENTIAL BIRTH AND DEATH PROCESS
'INTERVAL MATHEMATICS', ED. BY K. NICKEL
LECTURE NOTES IN COMPUTER SCIENCE 29, SPRINGER
VERLAG, 135-140 (1975)

48* AMES, W. F.; ADAMS, E.
MONOTONICALLY CONVERGENT TWO-SIDED BOUNDS FOR SOME
INVARIANT PARABOLIC BOUNDARY VALUE PROBLEMS
Z. ANGEW. MATH. MECH. 56, T240-T242 (1976)

49* AMES, W. F.; GINSBERG, M.
BILATERAL ALGORITHMS AND THEIR APPLICATIONS
TO APPEAR: PROCEEDINGS OF THE INTERNATIONAL
CONFERENCE OF COMPUTATIONAL METHODS IN NONLINEAR
MECHANICS (SPRINGER)

APOSTOLATOS, N.
SIEHE AUCH: ALEFELD, G.; APOSTOLATOS, N.
                      WIPPERMANN, H.-W. ET AL.

50* APOSTOLATOS, N.
NULLSTELLENBESTIMMUNG MIT FEHLERERFASSUNG
BER. D. RECHENZENTRUMS U. D. INST. F. ANGEW.
MATH., UNIVERSITAET KARLSRUHE (1967)

51* APOSTOLATOS, N.
EIN EINSCHLIESSUNGSVERFAHREN FUER NULLSTELLEN
(NOTIZ)
Z. ANGEW. MATH. MECH. 47, T80 (1967)

52* APOSTOLATOS, N.
ALLGEMEINE INTERVALLARITHMETIKEN UND ANWENDUNGEN
BULL. SOC. MATH. GRECE, (N. S.) 10, 135-180
(1969)

53* APOSTOLATOS, N.
EINE METHODE ZUR NULLSTELLENBESTIMMUNG VON
FUNKTIONEN
COMPUTING 15, 1-10 (1975)

54* APOSTOLATOS, N.
EINE ALLGEMEINE BETRACHTUNG VON NUMERISCHEN
ALGORITHMEN
INSTITUTSBERICHT NR. 119, (1975) TECHNISCHE
UNIVERSITAET MUENCHEN

55* APOSTOLATOS, N.
EIN ALGORITMUS ZUR NULLSTELLENBESTIMMUNG VON
FUNKTIONEN.
Z. ANGEW. MATH. MECH. 56, T273-T274 (1976)

56* APOSTOLATOS, N.; KULISCH, U.
GRUNDLAGEN EINER MASCHINENINTERVALLARITHMETIK
COMPUTING 2, 89-104 (1967)

57* APOSTOLATOS, N.; KULISCH, U.
APPROXIMATION DER ERWEITERTEN INTERVALLARITHMETIK
DURCH DIE EINFACHE MASCHINENINTERVALLARITHMETIK
COMPUTING 2, 181-194 (1967)

58* APOSTOLATOS, N.; KULISCH, U.
GRUNDZUEGE EINER INTERVALLRECHNUNG FUER MATRIZEN
UND EINIGE ANWENDUNGEN
ELEKTRON. RECHENANLAGEN 10, 73-83 (1968)

59* APOSTOLATOS, N.; KULISCH, U.; NICKEL, K.
EIN EINSCHLIESSUNGSVERFAHREN FUER NULLSTELLEN
COMPUTING 2, 195-201 (1967)

60* APPELT, W.
PRAECOMPILER FUER INTERVALLARITHMETIK
TAGUNG UEBER INTERVALLRECHNUNG IN OBERWOLFACH,
VORTRAGSMANUSKRIPT (1972)

APPELT-AVENHAUS

61* APPELT, W.
INTERVALLARITHMETISCHE SPLINEFUNKTIONEN
KURZMANUSKRIPT ZU EINEM VORTRAG IN BAD GODESBERG
(1972)

62* APPELT, W.
FEHLEREINSCHLIESSUNG BEI DER NUMERISCHEN LOESUNG
ELLIPTISCHER DIFFERENTIALGLEICHUNGEN UNTER
VERWENDUNG EINES INTERVALL-ARITHMETISCHEN
VERFAHRENS
BERICHTE DER GMD BONN 79 (1973)

63* APPELT, W.
FEHLEREINSCHLIESSUNG FUER DIE LOESUNG EINER KLASSE
ELLIPTISCHER RANDWERTAUFGABEN
Z. ANGEW. MATH. MECH. 54, T207-T208 (1974)

64* APPELT, W.
VERWENDUNG VON TSCHEBYSCHEW-APPROXIMATION UND
OEKONOMISIERUNGSVERFAHREN BEI DER VERGROEBERUNG
VON INTERVALLPOLYNOMEN
'INTERVAL MATHEMATICS', ED. BY K. NICKEL
LECTURE NOTES IN COMPUTER SCIENCE 29, SPRINGER
VERLAG, 141-149 (1975)

65* APPELT, W.
VERGROEBERUNG VON INTERVALLPOLYNOMEN
COMPUTING 15, 329-346 (1975)

66* APPELT, W.
VERGLEICH DER VERGROEBERUNGSVERFAHREN FUER
INTERVALLPOLYNOME VON KRUECKEBERG UND ROKNE
Z. ANGEW. MATH. MECH. 56, T274-T275 (1976)

67* APPELT, W.
EIN ABBRUCHKRITERIUM FUER DAS
INTERVALLARITHMETISCHE GESAMT- UND
EINZELSCHRITTVERFAHREN
ERSCHEINT DEMNAECHST

68* ARTHUR, D.W.
THE USE OF INTERVAL ARITHMETIC TO BOUND THE ZEROS
OF REAL POLYNOMIALS
J. INST. MATH. APPL. 10, 231-237 (1972)

ASHENHURST, R. L.
SIEHE AUCH: METROPOLIS, N.; ASHENHURST, R. L..

69* ASHENHURST, R. L.
NUMBER REPRESENTATION AND SIGNIFICANCE MONITORING
'MATHEMATICAL SOFTWARE', ED. BY J. R. RICE, NEW
YORK: ACADEMIC PRESS, 67-86 (1971)

70* ASHENHURST, R. L.; METROPOLIS, N.
SIGNIFICANT-DIGIT ARITHMETIC IN PRACTICE
PROC. IEEE NATIONAL ELECTRONICS CONFERENCE,
CHICAGO (1963)

71* ASHENHURST, R.L.; METROPOLIS, N.
ERROR ESTIMATION IN COMPUTER CALCULATION
AMER. MATH. MONTHLY 72, 47-58 (1965)

72* ATTEIA, M.
FONCTIONS 'SPLINE' AVEC CONTRAINTES LINEAIRES DE
TYPE INEGALITE
PROBLEMES DIFFERENTIELS ET INTEGRAUX, POUZET, M.P.
SESSION 12, 1.42-1.54 (1977)

AUBIN, J.-P.
SIEHE AUCH: CLARKE, F.H.; AUBIN, J.-P.

73* AVENHAUS, J.
ZUR NUMERISCHEN BEHANDLUNG DES ANFANGSWERTPROBLEMS
MIT EXAKTER FEHLERERFASSUNG
DISSERTATION, UNIVERSITAET KARLSRUHE (1970)

74* AVENHAUS, J.
EIN VERFAHREN ZUR EINSCHLIESSUNG DER LOESUNG DES
ANFANGSWERTPROBLEMS
COMPUTING 8, 182-190 (1971)

75* AVENHAUS, J.
ZUR NUMERISCHEN BEHANDLUNG DES ANFANGSWERTPROBLEMS
MIT EXAKTER FEHLERERFASSUNG (NOTIZ)
Z. ANGEW. MATH. MECH. 51, T34 (1971)

76* BACHMANN, K.-H.
UNTERSUCHUNGEN ZUR EINSCHLIESSUNG DER LOESUNGEN
VON SYSTEMEN GEWOEHNLICHER DIFFERENTIALGLEICHUNGEN
BEITR. Z. NUMER. MATH. 1, 9-42 (1974)

77* BACHMANN, K.-H.
EINSCHLIESSUNG DER NULLSTELLEN VON
INTERVALLPOLYNOMEN
BEITR. Z. NUMER. MATH. 4, 23-32 (1975)

78* BAHLKE, W.
EIN NUMERISCHER ALGORITHMUS ZUR BESTIMMUNG DER
MINIMALSTELLE UND DES MINIMALWERTES EINER STRENG
KONVEXEN FUNKTION MIT FEHLERSCHRANKEN
DIPLOMARBEIT AM INST. F. PRAKT. MATH.,
UNIVERSITAET KARLSRUHE (1973)

79* BALZER, H.-J,
DOPPELTGENAUE INTERVALLEXPONENTIALFUNKTION
INTERNER BER. NR. 8, INST. F. NUM.
DATENVERARBEITUNG, GMD BONN (1973)

80* BANKS, H.T.; JACOBS, M.Q.
A DIFFERENTIAL CALCULUS FOR MULTIFUNCTIONS
J. MATH. ANAL. APPL. 29, 246-272 (1970)

81* BARNSLEY, M.F.; ROBINSON, P.D.
BIVARIATIONAL BOUNDS ON SOLUTIONS OF TWO-POINT
BOUNDARY-VALUE PROBLEMS
J. MATH. ANAL. APPL. 56, 172-184 (1976)

82* BARTH, W.
NULLSTELLENBESTIMMUNG MIT DER INTERVALLRECHNUNG
COMPUTING 8, 320-328 (1971)

83* BARTH, W.
EIN ALGORITHMUS ZUR BERECHNUNG ALLER REELLEN
NULLSTELLEN IN EINEM INTERVALL
COMPUTING 9, 327-333 (1972)

84* BARTH, W.
ITERATIONSVERFAHREN DER INTERVALLRECHNUNG ZUR
LOESUNG LINEARER GLEICHUNGSSYSTEME
WIRD VEROEFFENTLICHT

85* BARTH, W.; NUDING, E.
OPTIMALE LOESUNG VON INTERVALLGLEICHUNGSSYSTEMEN
COMPUTING 12, 117-125 (1974)

86* BAUCH, H.
LOESUNGSEINSCHLIESSUNG BEI ANFANGSWERTAUFGABEN
GEWOEHNLICHER DIFFERENTIALGLEICHUNGEN MIT HILFE
INTERVALLANALYTISCHER METHODEN
DISSERTATION, UNIVERSITAET DRESDEN (1974)

87* BAUCH, H.
ZUR KONVERGENZ MONOTONER INTERVALLFOLGEN UND
MONOTONER FOLGEN INTERVALLWERTIGER FUNKTIONEN
Z. ANGEW. MATH. MECH. 55, 605-606 (1975)

88* BEECK, H.
UEBER INTERVALLANALYTISCHE METHODEN BEI LINEAREN
GLEICHUNGSSYSTEMEN MIT INTERVALLKOEFFIZIENTEN UND
ZUSAMMENHAENGE MIT DER FEHLERANALYSIS
DISSERTATION, TH MUENCHEN (1971)

89* BEECK, H.
UEBER STRUKTUR UND ABSCHAETZUNGEN DER
LOESUNGSMENGE VON LINEAREN GLEICHUNGSSYSTEMEN MIT
INTERVALLKOEFFIZIENTEN
COMPUTING 10, 231-244 (1972)

90* BEECK, H.
CHARAKTERISIERUNG DER LOESUNGSMENGE VON
INTERVALLGLEICHUNGSSYSTEMEN
Z. ANGEW. MATH. MECH. 53, T181-T182 (1973)

91* BEECK, H.
ZUR SCHARFEN AUSSENABSCHAETZUNG DER LOESUNGSMENGE
BEI LINEAREN INTERVALLGLEICHUNGSSYSTEMEN
Z. ANGEW. MATH. MECH. 54, T208-T209 (1974)

92* BEECK, H.
ZUR PROBLEMATIK DER HUELLENBESTIMMUNG VON
INTERVALL-GLEICHUNGSSYSTEMEN
'INTERVAL MATHEMATICS', ED. BY K. NICKEL
LECTURE NOTES IN COMPUTER SCIENCE 29, SPRINGER
VERLAG, 150-159 (1975)

93* BEECK, H.; FAHRMEIR, L.
BEMERKUNGEN ZUR BERUECKSICHTIGUNG VON TOLERANZEN
BEI INPUT-OUTPUT MODELLEN
ERSCHEINT DEMNAECHST

94* BELLINGHAUSEN, H.-K.
LOESUNG LINEARER GLEICHUNGSSYSTEME HOHER ORDNUNG
MITTELS ITERATION IN INTERVALLARITHMETIK
DIPLOMARBEIT AM INST. F. ANGEW. MATH. U.
INFORMATIK, UNIVERSITAET BONN (1970)

BEN-ISRAEL, A.
SIEHE AUCH: ROBERS, P.D.; BEN-ISRAEL, A.

95* BEN-ISRAEL, A.; ROBERS, P.D.
A DECOMPOSITION METHOD FOR INTERVAL LINEAR
PROGRAMMING
MANAGEMENT SCIENCE 16, 374-387 (1970)

96* BENNETT, G. K. JR.
A METHOD FOR LOCATING THE ZEROS OF A POLYNOMIAL
USING INTERVAL ARITHMETIC
COMPUTER CENTER, TEXAS TECHNOLOGICAL COLLEGE,
LUBBOCK, TEXAS (1967)

97* BERTI, S.N.
TEORIA RELATIILOR BINARE SI UNELE APLICATII
GAZ. MAT. SER. A. LXXIII, 428-434 (1968)

98* BERTI, S.N.
APLICATII ALE TEORIEI RELATIILOR IN LOGICA
MATEMATICA
STUD. CERC. MAT. 21, 3-11 (1969)

99* BERTI, S.N.
THE GEOMETRY OF RELATIONS IN THE SET OF REAL
NUMBERS
MATHEMATICA (CLUJ) 11 (34), 29-48 (1969)

100* BERTI, S.N.
THE SOLUTION OF AN INTERVAL EQUATION
MATHEMATICA (CLUJ) 11 (34), 189-194 (1969)

101* BERTI, S.N.
INTERVALELE SI ARITMETICA LOR IN ANALIZA NUMERICA
GAZ. MAT. SER. A. LXXV, 309-313 (1970)

102* BERTI, S.N.
ARITMETICA SI ANALIZA INTERVALELOR
REVISTA DE ANALIZA NUMERICA SI TEORIA APROXIMATIEI
1, 21-39 (1972)

103* BERTI, S.N.
SOME RELATIONS BETWEEN INTERVAL FUNCTIONS (I)
MATHEMATICA (CLUJ) 14 (37), 9-26 (1972)

104* BERTI, S.N.
ON THE INTERVAL EQUATION AX+B=CX+D
REV. ANAL. NUMER. THEORIE APPROXIMATION (CLUJ) 2,
11-26 (1973)

105* BERTI, S.N.
ASUPRA REZOLVARII UNUI SISTEM DE ECUATII BOOLEENE
REV. ANAL. NUMER. THEORIA APROXIMATIEI (CLUJ) 2,
31-44 (1973)

106* BERTI, S.N.
OPERATIILE 'OR' SI 'AND' DIN ARITMETICA
INTERVALELOR
REV. ANAL. NUMER. THEORIA APROXIMATIEI (CLUJ) 3,
31-34 (1974)

107* BERTI, S.N.
ON THE GENERALIZED SUM OF INTERVALS
REV. ANAL. NUMER. THEORIE APPROXIMATION (CLUJ) 3,
11-21 (1974)

108* BHATTACHARJEE, G. P.
SIEHE: MAJUMDER, K. L.; BHATTACHARJEE, G. P.

109* BIERBAUM, F.
INTERVALL-MATHEMATIK
EINE LITERATURUEBERSICHT
INTERNER BERICHT NR. 74/2 D. INST. F. PRAKT. MATH.
UNIVERSITAET KARLSRUHE (1974)

110* BIERBAUM, F.
EINSATZ DER INTERVALLARITHMETIK BEI DER
NUMERISCHEN KONVERGENZ VON ALGOL-60 PROGRAMMEN
'INTERVAL MATHEMATICS', ED. BY K. NICKEL
LECTURE NOTES IN COMPUTER SCIENCE 29, SPRINGER
VERLAG, 160-168 (1975)

111* BIERBAUM, F.
INTERVALL-MATHEMATIK
EINE LITERATURUEBERSICHT, NACHTRAG
INTERNER BERICHT NR. 75/3 D. INST. F. PRAKT. MATH.
UNIVERSITAET KARLSRUHE (1975)

112* BIERBAUM, F.
INTERVALL-MATHEMATIK
EINE LITERATURUEBERSICHT, 2. AUFLAGE
INTERNER BERICHT NR. 76/4 D. INST. F. PRAKT. MATH.
UNIVERSITAET KARLSRUHE (1976)

113* BINSTOCK, W.; HAWKES, J.; HSU, N.-T.
AN INTERVAL INPUT/OUTPUT PACKAGE FOR THE UNIVAC
1108
MRC TECHNICAL SUMMARY REPORT # 1212 UNIVERSITY OF
WISCONSIN, MADISON (1973)

114* BLUMHOFER, M.
UEBER EINIGE ALGORITHMEN ZUR NUMERISCHEN
EINSCHLIESSUNG DER LOESUNG BEI LINEAREN
GLEICHUNGSSYSTEMEN
DIPLOMARBEIT AM INST. F. ANGEW. MATH.,
UNIVERSITAET KARLSRUHE (1972)

115* BOCHE, R.E.
AN OPERATIONAL INTERVAL ARITHMETIC
IEEE NATIONAL ELECTRONICS CONFERENCE, CHICAGO
(1963)

116* BOCHE, R. E.
SOME OBSERVATIONS ON THE ECONOMICS OF INTERVAL
ARITHMETIC
COMM. ACM 8, 649 & 703 (1965)

117* BOCHE, R. E.
SPECIFICATIONS FOR AN INTERVAL INPUT PROGRAM
SUBMITTED TO ACM (1965)

118* BOCHE, R.E.
COMPLEX INTERVAL ARITHMETIC WITH SOME APPLICATIONS
TECHNICAL REPORT # 4-22-66-1 LOCKHEED MISSILES AND
SPACE CO., PALO ALTO, CAL. (1966)

119* BOCHE, R. E.
INTERVAL ARITHMETIC WITH SOME EXTENSIONS AND
APPLICATIONS
TEXAS TECHNOLOGICAL COLLEGE, LUBBOCK, TEXAS

120* BOCHE, R. E.
AN EFFICIENT DIFFERENTIAL EQUATIONS PROGRAM
TEXAS TECHNOLOGICAL COLLEGE, LUBBOCK, TEXAS

121* BOEHMER, K.; JACKSON, R.T.
A FORTRAN-TRIPLEX-PRE-COMPILER BASED ON THE
AUGMENT PRE-COMPILER
MRC TECHNICAL SUMMARY REPORT # 1732 UNIVERSITY OF
WISCONSIN, MADISON (1977)

BOGART, K.P.
SIEHE AUCH: TROTTER, W.T., JR.; BOGART, K.P.

BOHLENDER, G.
SIEHE AUCH: KULISCH, U.; BOHLENDER, G.

122* BOHLENDER, G.
GLEITKOMMABERECHNUNG REELLER FUNKTIONEN MIT
MAXIMALER GENAUIGKEIT
Z. ANGEW. MATH. MECH. 56, T275-T277 (1976)

123* BOHLENDER, G.
GENAUE SUMMATION VON GLEITKOMMAZAHLEN
COMPUTING SUPPLEMENTUM 1, ALBRECHT, R.;
KULISCH, U., SPRINGER-VERLAG WIEN, NEW YORK,
21-32 (1977)

124* BOHLENDER, G.
PRODUKTE UND WURZELN VON GLEITKOMMAZAHLEN
COMPUTING SUPPLEMENTUM 1, ALBRECHT, R.;
KULISCH, U., SPRINGER-VERLAG WIEN, NEW YORK,
33-46 (1977)

BOLAND, J. C.
SIEHE: LEKKERKERKER, G. G.; BOLAND, J. C.

125* BRAESS, D.
DIE KONSTRUKTION MONOTONER ITERATIONSFOLGEN ZUR
LOESUNGSEINSCHLIESSUNG BEI LINEAREN
GLEICHUNGSSYSTEMEN
ARCH. RAT. MECH. ANAL. 9, 97-106 (1962)

126* BRAESS, D.
MONOTONE ITERATIONSFOLGEN BEI GLEICHUNGSSYSTEMEN
MIT FEHLERHAFTEN KOEFFIZIENTEN UND
ITERATIONSBESCHLEUNIGUNG
NUMER. MATH. 7, 32-41 (1965)

127* BRAESS, D.
DIE BERECHNUNG DER FEHLERGRENZEN BEI LINEAREN
GLEICHUNGSSYSTEMEN MIT FEHLERHAFTEN KOEFFIZIENTEN
ARCH. RAT. MECH. ANAL. 19, 74-80 (1966)

128* BRAESS, D.; HADELER, K. P.
SIMULTANEOUS INCLUSION OF THE ZEROS OF A
POLYNOMIAL
NUMER. MATH. 21, 161-165 (1973)

129* BRAUN, J.A.; MOORE, R.E.
A PROGRAM FOR THE SOLUTION OF DIFFERENTIAL
EQUATIONS USING INTERVAL ARITHMETIC (DIFEQ) FOR
THE CDC 3600 AND THE CDC 1604
MRC TECHNICAL SUMMARY REPORT # 901 UNIVERSITY OF
WISCONSIN, MADISON (1968)

130* BREITER, M. C.; KELLER, C. L.; REEVES, T. E.
A PROGRAM FOR COMPUTING ERROR BOUNDS FOR THE
SOLUTION OF A SYSTEM OF DIFFERENTIAL EQUATIONS
REPORT ARL 69-0054 AEROSPACE RESEARCH LABS.,
WRIGHT-PATTERSON AIR FORCE BASE, OHIO (1969)

131* BRENT, R. P.
AN ALGORITHM WITH GUARANTEED CONVERGENCE FOR
FINDING A ZERO OF A FUNCTION
COMPUT. J. 14, 422-425 (1971)

132* BREUER, P.
RUNDUNGSFEHLER BEI DIREKTER AUFLOESUNG
GEODAETISCHER GLEICHUNGSSYSTEME
DEUTSCHE GEODAET. KOMMISSION BEI D. BAYER. AKAD.
D. WISS. REIHE C: DISSERTATIONEN HEFT NR. 135
(1969)

133* BRIGHT, H. S.; COLHOUN, B. A.; MALLORY, F. B.
A SOFTWARE SYSTEM FOR TRACING NUMERICAL
SIGNIFICANCE DURING COMPUTER PROGRAM EXECUTION
SJCC, 387-392 (1971)

134* BROCKHAUS, M.; ROTHMAIER, B.; SCHROTH, P.
     BENUTZERANLEITUNG FUER TRIPLEX-ALGOL IM SYSTEM
     HYDRA 2
     INTERNER BERICHT DES INST. F. INFORMATIK 69/11,
     UNIVERSITAET KARLSRUHE (1969)

135* BROSSART, F.-J.
     EINSCHLIESSUNG DES MINIMALINTERVALLS EINER
     KONVEXEN FUNKTION AUF EINEM INTERVALL AUS I(R)
     DIPLOMARBEIT AM INST. F. PRAKT. MATH.,
     UNIVERSITAET KARLSRUHE (1975)

136* BROWN, R. W.
     UPPER AND LOWER BOUNDS FOR SOLUTIONS OF INTEGRAL
     EQUATIONS
     'ERROR IN DIGITAL COMPUTATION II', ED. BY
     L.B.RALL, WILEY & SONS INC, NEW YORK, 219-230
     (1965)

137* BURKILL, J.C.
     FUNCTIONS OF INTERVALS
     PROC. LONDON MATH. SOC.   22, 275-336   (1924)

138* BURKILL, J.C.
     THE DERIVATIVES OF FUNCTIONS OF INTERVALS
     FUND. MATH. 5, 321-327 (1924)

139* BURSCHEID, H. J.
     ZUM BEGRIFF DER INTERVALLORDNUNG
     MATH. PHYS. SEMESTERBER. 20, 72-85 (1973)

140* CAPRANI, O.
     ROUNDOFF ERRORS IN FLOATING-POINT SUMMATION
     BIT 15, 5-9 (1975)

141* CAPRANI, O.; MADSEN, K.
     CONTRACTION MAPPINGS IN INTERVAL ANALYSIS
     BIT 15, 362-366 (1975)

142* CHARTRES, B.A.
     CONTROLLED PRECISION CALCULATIONS AND THE
     DANILEWSKI-METHOD
     REP. IBM-E-148/8, NSF-E-872, DIV. APPL. MATH.,
     BROWN UNIVERSITY, PROVIDENCE, R.I. (1964)

143* CHARTRES, B.A.
     AUTOMATIC CONTROLLED PRECISION CALCULATIONS
     J. ASSOC. COMPUT. MACH. 13, 386-403 (1966)

144* CHARTRES, B.A.; GEUDER, J.C.
     COMPUTABLE ERROR BOUNDS FOR DIRECT SOLUTION OF
     LINEAR EQUATIONS
     J. ASSOC. COMPUT. MACH. 14, 63-71 (1967)

145* CHRIST, H.
     PROZEDUREN FUER TRIPLEXRECHNUNGEN IN ALGOL X8
     RECHENZENTRUM, UNIVERSITAET KARLSRUHE (1967)

146* CHRIST, H.
     REALISIERUNG EINER MASCHINENINTERVALLARITHMETIK
     AUF BELIEBIGEN ALGOL-COMPILERN (NOTIZ)
     Z. ANGEW. MATH. MECH. 48, T110 (1968)

147* CHRIST, H.
     REALISIERUNG EINER MASCHINENINTERVALLARITHMETIK
     AUF BELIEBIGEN ALGOL-60 COMPILERN
     ELEKTRON. RECHENANLAGEN 10, 217-222 (1968)

148* CHUBA, W.; MILLER, W.
     QUADRATIC CONVERGENCE IN INTERVAL ARITHMETIC,
     PART I
     BIT 12, 284-290   (1972)

149* CLARKE, F.H.; AUBIN, J.-P.
     MONOTONE INVARIANT SOLUTIONS TO DIFFERENTIAL
     INCLUSIONS
     MRC TECHNICAL SUMMARY REPORT # 1729 UNIVERSITY OF
     WISCONSIN, MADISON (1977)

     COLHOUN, B. A.
     SIEHE: BRIGHT, H. S.; COLHOUN, B. A.;
            MALLORY, F. B.

150* COLLATZ, L.
APPROXIMATION METHODS FOR EXPANDING OPERATORS
NUMER. ANAL., PROC. DUNDEE CONFERENCE 1975 LECTURE
NOTES MATH. 506, 49–59 (1976)

151* COLLINS, G.
INTERVAL ARITHMETIC FOR AUTOMATIC ERROR ANALYSIS
REPORT M&A– 5, MATHEMATICS AND APPLICATIONS DEP.,
IBM NEW YORK (1960)

MC CORMICK, G.P.
SIEHE AUCH: MANCINI, L.J.; MC CORMICK, G.P.

152* CRANE, M.A.
A BOUNDING TECHNIQUE FOR POLYNOMIAL FUNCTIONS
SIAM J. APPL. MATH. 29, 751–754 (1975)

153* CRARY, F.D.; LADNER, T.D.
A SIMPLE METHOD OF ADDING A NEW DATA TYPE TO
FORTRAN
MRC TECHNICAL SUMMARY REPORT # 1065 UNIVERSITY OF
WISCONSIN, MADISON (1970)

154* CRYER, C.W.
ON THE COMPUTATION OF RIGOROUS ERROR BOUNDS FOR
THE SOLUTIONS OF LINEAR EQUATIONS WITH THE AID OF
INTERVAL ARITHMETIC
COMPUTER SCIENCE TECHNICAL REPORT # 70 UNIVERSITY
OF WISCONSIN, MADISON (1969)

155* CSAJKA, I.; MUENZER, W.; NICKEL, K.
SUBROUTINES ADD, NEG, SUB, DIV, MUL, FOR USE IN AN
ERROR–BOUND ARITHMETIC
IBM RESEARCH LABORATORY, RUESCHLIKON (SCHWEIZ)

156* DANIEL, J.W.; MOORE, R.E.
COMPUTATION AND THEORY IN ORDINARY DIFFERENTIAL
EQUATIONS
W.H. FREEMAN, SAN FRANCISCO, (1970)

157* DARGEL, R.H.; LOSCALZO, F.R.; WITT, T.H.
INTERVAL ROOTFINDING PACKAGE (ROOTFIND) MANUSKRIPT
(1965)

158* DARGEL,R.H.; LOSCALZO,F.R.; WITT,T.H.
AUTOMATIC ERROR BOUNDS ON REAL ZEROS OF RATIONAL
FUNCTIONS
COMM. ACM 9, 806–809 (1966)

159* DAVEY, D.P.
GUARANTEED BOUNDS ON THE NUMERICAL SOLUTIONS OF
INITIAL VALUE PROBLEMS USING POLYTOPE ARITHMETIC
DISSERTATION, UNIVERSITY OF MONTREAL (1972)

160* DAVEY, D. P.; STEWART, N. F.
GUARANTEED ERROR BOUNDS FOR THE INITIAL VALUE
PROBLEM USING POLYTOPE ARITHMETIC
PUBLICATION # 109, DEPARTEMENT D'INFORMATIQUE,
UNIVERSITE DE MONTREAL (1972) PUBLIZIERT IN: BIT
16, 257–268 (1976)

DAVIDSON, J. A.
SIEHE: MOORE, R.E.; DAVIDSON, J.A.;
        JASCHKE, H.R.; SHAYER, S.

161* DAVIS, P.; RABINOWITZ, P.
ON THE ESTIMATION OF QUADRATURE ERRORS FOR
ANALYTIC FUNCTIONS
MATH. TABLES AND OTHER AIDS TO COMP. 8, 193–203
(1954)

162* DEGELOW, L.
UEBER DIE GUENSTIGSTE SCHRITTWEITE BEI
QUADRATURVERFAHREN
DIPLOMARBEIT AM INST. F. PRAKT. MATH.,
UNIVERSITAET KARLSRUHE (1973)

163* DEMPSTER, M.
DISTRIBUTIONS IN INTERVALS AND LINEAR PROGRAMMING
"TOPICS IN INTERVAL ANALYSIS", ED. BY E.HANSEN,
OXFORD UNIVERSITY PRESS, 107–127 (1969)

164* DEWAR, J. K. S.
PROCEDURES FOR INTERVAL ARITHMETIC
COMPUT. J. 14, 447-450 (1971)

165* DITTRICH, E.
NUMERISCHE BESTIMMUNG DER NULLSTELLEN DER GANZEN
FUNKTION F(Z) = INTEGRAL VON (BIS Z UEBER
EXP(-T POWER 2)
STAATSEXAMENSARBEIT, ANGEFERTIGT AM INST. F.
PRAKTISCHE MATH., UNIV. KARLSRUHE, (1976)

166* DORNBUSCH, W.
ZUR INTERVALL-ANALYTISCHEN FEHLERERFASSUNG BEI DER
NUMERISCHEN BEHANDLUNG VON INTEGRALGLEICHUNGEN
DIPLOMARBEIT AM INST. F. ANGEW. MATH. U.
INFORMATIK, UNIVERSITAET BONN (1971)

167* DOTTERWEICH, J.
MULTIVALLARITHMETIK EINE ARITHMETIK FUER
INTERVALLE MIT WAHRSCHEINLICHKEITSVERTEILUNGEN
DIPLOMARBEIT AM INST. F. ANGEW. MATH. U.
INFORMATIK, UNIVERSITAET BONN (1973)

DRESSEL, H.
SIEHE AUCH: SCHMIDT, J.W.; DRESSEL, H.

168* DUSSEL, R.
EINSCHLIESSUNG DES MINIMALPUNKTES EINER STRENG
KONVEXEN FFUNKTION AUF EINEM N-DIMENSIONALEN QUADER
DISSERTATION, INTERNER BERICHT DES INST. F. PRAKT.
MATH. 72/2, UNIVERSITAET KARLSRUHE (1972)

169* DUSSEL, R.
EINSCHLIESSUNG DES MINIMALPUNKTES EINER STRENG
KONVEXEN FUNKTION PHI AUF EINEM N-DIMENSIONALEN
QUADER Q (NOTIZ)
Z. ANGEW. MATH. MECH. 53, T184 (1973)

170* DUSSEL, R.
EINSCHLIESSUNG DES MINIMALPUNKTES EINER STRENG
KONVEXEN FUNKTION AUF EINEM N-DIMENSIONALEN QUADER
'INTERVAL MATHEMATICS', ED. BY K. NICKEL

LECTURE NOTES IN COMPUTER SCIENCE 29, SPRINGER
VERLAG, 169-177 (1975)

171* DUSSEL, R.; SCHMITT, B.
DIE BERECHNUNG VON SCHRANKEN FUER DEN WERTEBEREICH
EINES POLYNOMS IN EINEM INTERVALL
INTERNER BERICHT DES INST. F. INFORMATIK 69/2,
UNIVERSITAET KARLSRUHE (1969)
PUBLIZIERT IN:
COMPUTING 6, 35-60 (1970)

172* DUSSEL, R.; SCHMITT, B.
DIE BERECHNUNG VON SCHRANKEN FUER DEN WERTEBEREICH
EINES POLYNOMS IN EINEM INTERVALL (NOTIZ)
Z. ANGEW. MATH. MECH. 50, T80 (1970)

173* DWYER, P.S.
COMPUTATION WITH APPROXIMATE NUMBERS
'LINEAR COMPUTATIONS', ED. BY P.S. DWYER, WILEY &
SONS INC, NEW YORK, 11-35 (1951)

174* DWYER, P.S.
MATRIX INVERSION WITH THE SQUARE ROOT METHOD
TECHNOMETRICS 6, 197-213 (1964)

175* DYCK, H.-P.
ALGORITHMEN ZUR NUMERISCHEN BEHANDLUNG VON
FUNKTIONEN MIT FEHLERERFASSUNG
4973.64192591 14 95ST# 6# 1575W# 41T8#,
DIPLOMARBEIT AM INST. F. ANGEW. MATH.,

176* ECKER, K.; RATSCHEK, H.
INTERVALLARITHMETIK MIT
WAHRSCHEINLICHKEITSSTRUKTUR
ANGEW. INFORMATIK (ELEKTRON. DATENVERARBEITUNG)
14, 313-320 (1972)

177* EIJGENRAAM, P.; GRIEND, J.A.VAN DE;
STATEMA, L.S.C.
TRIPLEX-ALGOL 60 ON THE IBM 370/158 OF THE
CENTRAAL REKENINSTITUUT
DIPLOMARBEIT, UNIVERSITY OF LEIDEN (1976)

EILENBERG-FISCHER

178* EILENBERG, S.; MONTGOMERY, D.
FIXED POINT THEOREMS FOR MULTIVALUED
TRANSFORMATIONS
AMER. J. MATH. 68, 214-222 (1946)

179* ELLER, K.
LOCAL ERROR BOUNDS IN THE NUMERICAL SOLUTION OF
ORDINARY DIFFERENTIAL EQUATIONS
MASTER'S THESIS, UNIVERSITY OF GUELPH (1970)

180* FAASS, E.
SIEHE AUCH: ADAMS, E.; FAASS, E.

181* FAASS, E.
BELIEBIG GENAUE NUMERISCHE SCHRANKEN FUER DIE
LOESUNG PARABOLISCHER RANDWERTAUFGABEN
'INTERVAL MATHEMATICS', ED. BY K. NICKEL
LECTURE NOTES IN COMPUTER SCIENCE 29, SPRINGER
VERLAG, 178-183 (1975)

182* FAASS, E.
BELIEBIG GENAUE NUMERISCHE SCHRANKEN FUER DIE
LOESUNG PARABOLISCHER RANDWERTAUFGABEN
DISSERTATION, UNIVERSITAET KARLSRUHE (1975)

183* FAASS, E.; ADAMS, E.
KONSTRUKTION STRENG GUELTIGER SCHRANKEN FUER EIN
SYSTEM PARABOLISCHER DIFFERENTIALGLEICHUNGEN
Z. ANGEW. MATH. MECH. 52, T182-T184 (1972)

FAHRMEIR, L.
SIEHE: BEECK, H.; FAHRMEIR, L.

184* FISCHER, H.
EINE METHODE ZUR EINSCHLIESSUNG DER NULLSTELLEN
VON POLYNOMEN MIT KOMPLEXEN KOEFFIZIENTEN
COMPUTING 4, 125-138 (1969)

185* FISCHER, H.
INTERVALL-ARITHMETIKEN FUER KOMPLEXE ZAHLEN
Z. ANGEW. MATH. MECH. 53, T190-T191 (1973)

186* FISCHER, H.
INTERVALLRECHNUNG UND HYPERNORMBALL-RECHNUNG IM CN
THE GREEK MATH. SOCIETY ED. BY THE GREEK MATH.
SOC., ATHEN. 136-145 (1974)

187* FISCHER, H.
HYPERNORMBAELLE ALS ABSTRAKTE SCHRANKENZAHLEN
COMPUTING 12, 67-73 (1974)

188* FISCHER, H.
UEBER VEKTORIELLE NORMEN UND DEN ZUSAMMENHANG MIT
PSEUDOMETRIKEN
Z. ANGEW. MATH. MECH. 54, T188-T189 (1974)

189* FISCHER, H.
NORMBAELLE IN DER INTERVALLRECHNUNG
Z. ANGEW. MATH. MECH. 54, T217-T218 (1974)

190* FISCHER, H.
EIN VERSUCH ZUR VERALLGEMEINERUNG DER
INTERVALLRECHNUNG
'INTERVAL MATHEMATICS', ED. BY K. NICKEL
LECTURE NOTES IN COMPUTER SCIENCE 29, SPRINGER
VERLAG, 184-190 (1975)

191* FISCHER, H.
INTERVALL-OPERATIONEN
INSTITUTSBERICHT NR. 116 (1975)
INSTITUT FUER STATISTIK UND UNTERNEHMENSFORSCHUNG,
TECHNISCHE UNIVERSITAET MUENCHEN

192* FISCHER, H.; OST, F.
ZUR THEORIE DER HYPERNORMBAELLE IN LINEAREN
RAEUMEN
INSTITUTSBERICHT NR. 114 (1975)
INSTITUT FUER STATISTIK UND UNTERNEHMENSFORSCHUNG,
TECHNISCHE UNIVERSITAET MUENCHEN

193* FISCHER, H.; OST, F.
GITTEREIGENSCHAFTEN VON HNB-MENGEN
INSTITUTSBERICHT NR. 118 (1975)
INSTITUT FUER STATISTIK UND UNTERNEHMENSFORSCHUNG.
TECHNISCHE UNIVERSITAET MUENCHEN

FISCHER-GARGANTINI

194* FISCHER, H.; OST, F.
GITTEREIGENSCHAFTEN VON TOLERANZ-UMGEBUNGEN
Z. ANGEW. MATH. MECH. 57, T275-T277 (1977)

195* FISCHER, H.; SURKAN, A.J.
IMPLEMENTATION OF INTERVAL ARITHMETIC USING APL
KEIN LITERATURNACHWEIS

196* FISCHER, P. C.
AUTOMATIC PROPAGATED AND ROUND-OFF ERROR ANALYSIS
13. NATIONAL MEETING OF THE ACM 1958 (1958)

197* FISHBURN, P. C.
INTRANSITIVE INDIFFERENCE WITH UNEQUAL
UNINDIFFERENCE INTERVALS
J. MATHEMATICAL PSYCHOLOGY 7, 144-149 (1970)

198* FISHBURN, P. C.
BETWEENNESS, ORDERS AND INTERVAL GRAPHS
J. PURE APPL ALGEBRA 1, 159-178 (1971)

199* FON ROBUE
AN INTERVAL ANALYSIS APPROACH TO THE STOCK MARKET
J. OF IRREPRODUCIBLE RESULTS 2, 24 (1975)

200* FRANZEN, R.
DIE INTERVALLANALYTISCHE BEHANDLUNG DES
EIGENWERTPROBLEMS VON MATRIZEN
DIPLOMARBEIT AM INST. F. ANGEW. MATH. U.
INFORMATIK, UNIVERSITAET BONN (1968)

201* FRANZEN, R.
DIE INTERVALLANALYTISCHE BEHANDLUNG
PARAMETERABHAENGIGER GLEICHUNGSSYSTEME
BERICHTE DER GMD BONN 47 (1971)

202* FRANZEN, R.
DIE KONSTRUKTION EINES APPROXIMATIONSPOLYNOMS FUER
DIE LOESUNGEN PARAMETERABHAENGIGER
GLEICHUNGSSYSTEME
Z. ANGEW. MATH. MECH. 52, T202-T204 (1972)

    GACHES, J.
    SIEHE AUCH: RIGAL, J.L.; GACHES, J.

203* GACHES, M.
COMPATIBILITE D'UNE SOLUTION CALCULEE AVEC LES
DONNES D'UN SYSTEME LINEAIRE
PROGRAMMATION EN MATHEMATIQUE NUMERIQUES (COLLOQUE
BESANCON 1966), EDITIONS DU CNRS, PARIS, 135-140
(1968)

204* GALPERIN, E. A.
THE CONDITION PROBLEM IN SOLUTION OF LINEAR
MULTISTAGE SYSTEMS
'INTERVAL MATHEMATICS', ED. BY K. NICKEL
LECTURE NOTES IN COMPUTER SCIENCE 29, SPRINGER
VERLAG, 191-195 (1975)

205* GANZ, H.
VERFAHREN ZUR EINSCHLIESSUNG DER LOESUNG EINER
ANFANGSWERTAUFGABE
DIPLOMARBEIT AM INST. F. PRAKT. MATH.,
UNIVERSITAET KARLSRUHE (1975)

206* GARDINER, V.; METROPOLIS, W.
A COMPREHENSIVE APPROACH TO COMPUTER ARITHMETIC
LA-4531, LOS ALAMOS SCIENTIFIC LABORATORY (1970)

    GARGANTINI, I.
    SIEHE AUCH: HENRICI, P.; GARGANTINI, I.

207* GARGANTINI, I.
PARALLEL ALGORITHMS FOR THE DETERMINATION OF
POLYNOMIAL ZEROS
PROC. 3RD. MANITOBA CONF. NUMER MATH., WINNIPEG
(1973) 195-211 (1974)

208* GARGANTINI, I.
PARALLEL SQUARE ROOT ITERATIONS
'INTERVAL MATHEMATICS', ED. BY K. NICKEL
LECTURE NOTES IN COMPUTER SCIENCE 29, SPRINGER
VERLAG, 196-204 (1975)

GARGANTINI–GLATZ

209* GARGANTINI, I.
COMPARING PARALLEL NEWTON'S METHOD WITH PARALLEL
LAGUERRE'S METHOD
VORTRAG IN OBERWOLFACH (1976)
INTERVALL-MATHEMATIK-TAGUNG
PUBLIZIERT IN: COMP. AND MATH. WITH APPLS. 2,
201-206 (1976)

210* GARGANTINI, I.
PARALLEL LAGUERRE ITERATIONS: THE COMPLEX CASE
NUMER. MATH. 26, 317-323 (1976)

211* GARGANTINI, I.; HENRICI, P.
CIRCULAR ARITHMETIC AND THE DETERMINATION OF
POLYNOMIAL ZEROS
NUMER. MATH. 18, 305-320 (1972)

212* GARLOFF, J.
EIN VERFAHREN ZUR BESTIMMUNG DER INNEREN LOESUNGEN
VON INTERVALL-GLEICHUNGSSYSTEMEN
DIPLOMARBEIT AM INST. F. ANGEW. MATH.,
UNIVERSITAET HEIDELBERG (1976)

213* GEGUSCH, T.
EINE DEZIMALE TRIPLEX-LANGZAHL-ARITHMETIK FUER DEN
UNIVAC-ALGOL-60-COMPILER
DIPLOMARBEIT AM INST. F. PRAKT. MATH.,
UNIVERSITAET KARLSRUHE (1976)

GEUDER, J.C.
SIEHE: CHARTRES, B.A.; GEUDER, J.C.

214* GEUL, A.
ALGEBRAISCHE EIGENSCHAFTEN DER
INTERVALLARITHMETIK,
DIE QUASILINEAREN RAEUME DER INTERVALLE,
INTERVALLVEKTOREN UND INTERVALLMATRIZEN UND IHRE
ANWENDUNG AUF PROBLEME DER LINEAREN ALGEBRA
DIPLOMARBEIT AM INSTITUT FUER INFORMATIK,
UNIVERSITAET STUTTGART (1972)

215* GIBB, A.
PROCEDURES FOR RANGE ARITHMETIC (ALGORITHM 61)
COMM. ACM 4, 319-320 (1961)

216* GIBB, A.
ALGOL PROCEDURES FOR RANGE ARITHMETIC
TECHNICAL REPORT 15, APPLIED MATHEMATICS AND
STATISTICS LABORATORIES, STANFORD UNIVERSITY
CALIFORNIA (1961)

217* GIBB, A.
A POSTERIORI BOUNDS ON A ZERO OF A POLYNOMIAL
(ALGORITHM 174)
COMM. ACM 6, 311-312 (1963)

GINSBERG, M.
SIEHE AUCH: AMES, W. F.; GINSBERG, M.

218* GINSBERG, M.
BILATERAL ALGORITHMS FOR CERTAIN CLASSES OF
ORDINARY DIFFERENTIAL EQUATIONS
DOCTORAL DISSERTATION, UNIVERSITY OF IOWA, IOWA
CITY (1972)

219* GLATZ, G.
DIE NUMERISCHE BERECHNUNG VON FEHLERSCHRANKEN ZU
NAEHERUNGSWERTEN VON POLYNOMWURZELN
DIPLOMARBEIT AM INSTITUT FUER INFORMATIK,
UNIVERSITAET KARLSRUHE (1969)

220* GLATZ, G.
NEWTON-ALGORITHMEN ZUR BESTIMMUNG VON
POLYNOMWURZELN UNTER VERWENDUNG KOMPLEXER
KREISARITHMETIK
"INTERVAL MATHEMATICS", ED. BY K. NICKEL
LECTURE NOTES IN COMPUTER SCIENCE 29, SPRINGER
VERLAG, 205-214 (1975)

221* GLATZ, G.
A-POSTERIORI-FEHLERSCHRANKEN BEI DER NUMERISCHEN
BERECHNUNG INSBESONDERE SCHLECHT KONDITIONIERTER
POLYNOMNULLSTELLEN
DISSERTATION, INTERNER BERICHT D. INST. F. PRAKT.
MATH. 75/7, UNIVERSITAET KARLSRUHE (1975)

222* GLATZ, G.
A-POSTERIORI-FEHLERSCHRANKEN INSBESONDERE SCHLECHT
KONDITIONIERTER POLYNOMWURZELN
Z. ANGEW. MATH. MECH. 56, T288-T289 (1976)

GOLDSTEIN, A.J.
SIEHE: RICHMAN, P. L.; GOLDSTEIN, A. J.

223* GOLDSTEIN, M.
SIGNIFICANCE ARITHMETIC ON A DIGITAL COMPUTER
COMM. ACM 6, 111-117 (1963)

224* GOLDSTEIN, M.; HOFFBERG, S.
THE ESTIMATION OF SIGNIFICANCE
"MATHEMATICAL SOFTWARE", ED. BY J. R. RICE, NEW
YORK: ACADEMIC PRESS, 93-104 (1971)

225* GOOD, D.I.; LONDON, R.L.
INTERVAL ARITHMETIC FOR THE BORROUGHS B-5500 FOUR
ALGOL-PROCEDURES AND PROOFS OF THEIR CORRECTNESS
MRC TECHNICAL SUMMARY REPORT # 26 UNIVERSITY OF
WISCONSIN, MADISON (1968)

226* GOOD, D.I.; LONDON, R.L.
COMPUTER INTERVAL ARITHMETIC: DEFINITION AND PROOF
OF CORRECT IMPLEMENTATION
J. ASSOC. COMPUT. MACH. 17, 603-612 (1970)

227* GORN, S.
ON THE STUDY OF COMPUTATIONAL ERRORS BALLISTIC
RESEARCH LABORATORIES REPORT 816
ABERDEEN PROVING GROUND, MARYLAND (1952)

228* GORN, S.
THE AUTOMATIC ANALYSIS AND CONTROL OF COMPUTING
ERRORS
J. SOC. INDUST. MATH. 2, 69-81 (1954)

229* GORN, S.; MOORE, R.E.
AUTOMATIC ERROR CONTROL - THE INITIAL VALUE
PROBLEM IN ORDINARY DIFFERENTIAL EQUATIONS
BALLISTIC RESEARCH LABORATORIES REPORT 893
ABERDEEN PROVING GROUND, MARYLAND (1953)

230* GRANT, J.A.; HITCHINS, G.D.
THE SOLUTION OF POLYNOMIAL EQUATIONS IN INTERVAL
ARITHMETIC
COMPUT. J. 16, 69-72 (1973)

231* GRAY, H. L.; HARRISON, C. JR.
NORMALIZED FLOATING-POINT ARITHMETIC WITH AN INDEX
OF SIGNIFICANCE
PJCC 16, 244-248 (1959)

232* GRAY, J.H.; RALL, L.B.
NEWTON: A GENERAL PURPOSE PROGRAM FOR SOLVING
NONLINEAR SYSTEMS
MRC TECHNICAL SUMMARY REPORT # 790 UNIVERSITY OF
WISCONSIN, MADISON (1967)

233* GRAY, J.H.; RALL, L.B.
INTE: A UNIVAC 1108/1110 PROGRAM FOR NUMERICAL
INTEGRATION WITH RIGOROUS ERROR ESTIMATION
MRC TECHNICAL SUMMARY REPORT # 1428 UNIVERSITY OF
WISCONSIN, MADISON (1975)

234* GREENSPAN, D.
INTERVAL ARITHMETIC
"INTRODUCTION TO NUMERICAL ANALYSIS AND
APPLICATIONS", ED. BY D. GREENSPAN, MARKHAM PUBL.
CO., CHICAGO, 147-165 (1971)

GRIEND, J.A.VAN DE
SIEHE AUCH: EIJGENRAAM, P.; GRIEND, J.A.VAN DE;
STATEMA, L.S.C.

GROSS-HAHN

GROSS, A.
SIEHE: FULKERSON, D.; GROSS, A.

235* GROSS, H.
SCHRANKEN FUER DIE EIGENWERTE EINER SCHWINGENDEN
MEMBRAN AUF EINEM L-FOERMIGEN GEBIET
DIPLOMARBEIT AM INST. F. PRAKT. MATH.,
UNIVERSITAET KARLSRUHE (1974)

236* GRUENER, K.
AUFLOESUNG LINEARER GLEICHUNGSSYSTEME MIT
ERHOEHTER GENAUIGKEIT
Z. ANGEW. MATH. MECH. 56, T290-T292 (1976)

237* GRUENER, K.
FEHLERSCHRANKEN FUER LINEARE GLEICHUNGSSYSTEME
COMPUTING SUPPLEMENTUM 1, ALBRECHT, R.;
KULISCH, U.; SPRINGER-VERLAG WIEN, NEW YORK,
47-55 (1977)

GRUETZMANN, J.
SIEHE AUCH: WALLISCH, W.; GRUETZMANN, J.

238* GUDERLEY, K. G.; KELLER, C. L.
ELLIPSOIDAL BOUNDS FOR THE SOLUTIONS OF SYSTEMS OF
ORDINARY LINEAR DIFFERENTIAL EQUATIONS
REPORT ARL 69-0005 AEROSPACE RESEARCH LABS.;
WRIGHT-PATTERSON AIR FORCE BASE, OHIO (1969)

239* GUDERLEY, K. G.; KELLER, C. L.
ELLIPSOIDAL BOUNDS FOR THE VECTOR SUM OF TWO
ELLIPSOIDS
REPORT ARL 69-0017 AEROSPACE RESEARCH LABS.,
WRIGHT-PATTERSON AIR FORCE BASE, OHIO (1969)

240* GUDERLEY, K. G.; KELLER, C. L.
ENCLOSING THE VECTOR SUM OF TWO ELLIPSOIDS WITH AN
ELLIPSOID
ABH. MATH. SEM. UNIV. HAMBURG 36, 173-184 (1971)

241* GUDERLEY, K. G.; KELLER, C. L.
A BASIC THEOREM IN THE COMPUTATION OF ELLIPSOIDAL
ERROR BOUNDS
NUMER. MATH. 19, 218-229 (1972)

242* GUDERLEY, K. G.; VALENTINE, M.
ON ERROR BOUNDS FOR THE SOLUTION OF SYSTEMS OF
ORDINARY DIFFERENTIAL EQUATIONS
BLANCH ANNIVERSARY VOLUME, AEROSPACE RESEARCH
LABS., WRIGHT-PATTERSON AIR FORCE BASE, OHIO,
45-89 (1967)

243* GUTKNECHT, M.
FEHLERFORTPFLANZUNG BEI DER BERECHNUNG DER
NULLSTELLEN EINES POLYNOMS DURCH SUKZESSIVES
ABSPALTEN VON LINEARFAKTOREN
DIPLOMARBEIT, ETH ZUERICH (1969)

244* GUTKNECHT, M.
A PRIORI FEHLERSCHRANKEN FUER SUKZESSIV
ABGESPALTENE POLYNOMNULLSTELLEN
Z. ANGEW. MATH. PHYS. 22, 630-634 (1971)

245* GUTKNECHT, M.
A POSTERIORI ERROR BOUNDS FOR THE ZEROS OF A
POLNOMIAL
NUMER. MATH. 20, 139-148 (1972)

HADELER, K. P.
SIEHE: BRAESS, D; HADELER, K. P.

246* HAERTRICH, F.
DIREKTE KONSTRUKTION VON POLYNOMSTREIFEN FUER
STANDARDFUNKTIONEN
DIPLOMARBEIT AM INST. F. ANGEW. MATH. U.
INFORMATIK, UNIVERSITAET BONN (1970)

247* HAHN, W.
INTERVALLARITHMETIK IN NORMIERTEN RAEUMEN UND
ALGEBREN
DISSERTATION, BERICHT DES INST. F. ANGEW. MATH. 3,
UNIVERSITAET GRAZ (1971)

HANSEN-HAUENSCHILD

248* HANSEN, E.
INTERVAL ARITHMETIC IN MATRIX COMPUTATIONS, PART I
SIAM J. NUMER. ANAL. 2, 308-320 (1965)

249* HANSEN, E.
ON SOLVING SYSTEMS OF EQUATIONS USING INTERVAL
ARITHMETIC
MATH. COMP. 22, 374-384 (1968)

250* HANSEN, E.
ON THE SOLUTION OF LINEAR ALGEBRAIC EQUATIONS WITH
INTERVAL COEFFICIENTS
LINEAR ALGEBRA AND APPL. 2, 153-165 (1969)

251* HANSEN, E.
TOPICS IN INTERVAL ANALYSIS
OXFORD UNIVERSITY PRESS (1969)

252* HANSEN, E.
ON LINEAR ALGEBRAIC EQUATIONS WITH INTERVAL
COEFFICIENTS
'TOPICS IN INTERVAL ANALYSIS', ED. BY E. HANSEN,
OXFORD UNIVERSITY PRESS, 35-46 (1969)

253* HANSEN, E.
ON SOLVING TWO-POINT BOUNDARY-VALUE PROBLEMS USING
INTERVAL ARITHMETIC
'TOPICS IN INTERVAL ANALYSIS', ED. BY E. HANSEN,
OXFORD UNIVERSITY PRESS, 74-90 (1969)

254* HANSEN, E.
THE CENTERED FORM
'TOPICS IN INTERVAL ANALYSIS', ED. BY E. HANSEN,
OXFORD UNIVERSITY PRESS, 102-106 (1969)

255* HANSEN, E.
A GENERALIZED INTERVAL ARITHMETIC
'INTERVAL MATHEMATICS', ED. BY K. NICKEL
LECTURE NOTES IN COMPUTER SCIENCE 29, SPRINGER
VERLAG, 7-18 (1975)

256* HANSEN, E.; SMITH, R.
A COMPUTER PROGRAM FOR SOLVING A SYSTEM OF LINEAR
EQUATIONS AND MATRIX INVERSION WITH AUTOMATIC
ERROR BOUNDING USING INTERVAL ARITHMETIC
REPORT LMSC 4-22-66-3, LOCKHEED MISSILES AND SPACE
CO., PALO ALTO, CAL. (1966)

257* HANSEN, E.; SMITH, R.
INTERVAL ARITHMETIC IN MATRIX COMPUTATIONS,
PART II
SIAM J. NUMER. ANAL. 4, 1-9 (1967)

258* HANSON, R. J.
INTERVAL ARITHMETIC AS A CLOSED ARITHMETIC SYSTEM
ON A COMPUTER
JPL SECTION 314, TECHNICAL MEMORANDUM 197 (1968)

259* HANSON, R. J.
AUTOMATIC ERROR BOUNDS FOR REAL ROOTS OF
POLYNOMIALS HAVING INTERVAL COEFFICIENTS
COMPUT. J. 13, 284-288 (1970)

HARRISON, C. JR.
SIEHE: GRAY, H. L.; HARRISON, C. JR.

260* HARTEN, U.
EIN VERFAHREN ZUR VOLLAUTOMATISCHEN BERECHNUNG DER
SCHRANKEN FUER DIE EXTREMWERTE EINES POLYNOMS P(X)
IN EINEM ABGESCHLOSSENEN INTERVALL
DIPLOMARBEIT AM INST. F. INFORMATIK, UNIVERSITAET
KARLSRUHE (1969)

261* HAUENSCHILD, M.
ANSAETZE ZUR KOMPLEXEN KREISARITHMETIK
ARBEITSBERICHT DES RECHENZENTRUMS NR. 7304,
UNIVERSITAET BOCHUM (1973)

262* HAUENSCHILD, M.
UEBER KREISSCHEIBENARITHMETIK UND KOMPLEXE
INTERVALLRECHNUNG
DIPLOMARBEIT, UNIVERSITAET BOCHUM (1973)
MANUSKRIPT (1973)

HAUENSCHILD–HENRICI

263* HAUENSCHILD, M.
ARITHMETIKEN FUER KOMPLEXE KREISE
COMPUTING 13, 299-312 (1974)

HAWKES, J.
SIEHE: BINSTOCK, W.; HAWKES, J.; HSU, N.-T.

264* HEBGEN, M.
DIE PIVOTSUCHE BEI REELLER UND
INTERVALLARITHMETISCHER LOESUNG VON LINEAREN
GLEICHUNGSSYSTEMEN DURCH ELIMINATIONSVERFAHREN
DIPLOMARBEIT, INST. F. ANGEW. MATH., UNIVERSITAET
HEIDELBERG (1973)

265* HEBGEN, M.
EINE SCALING-INVARIANTE PIVOTSUCHE FUER
INTERVALLMATRIZEN
COMPUTING 12, 99-106 (1974)

266* HEBGEN, M.
EIN ITERATIONSVERFAHREN, WELCHES DIE OPTIMALE
INTERVALLEINSCHLIESSUNG DER INVERSEN EINES
M-MATRIXINTERVALLS LIEFERT
COMPUTING 12, 107-115 (1974)

267* HEIDT, M.
ZUR NUMERISCHEN LOESUNG VON GEWOEHNLICHEN
DIFFERENTIALGLEICHUNGEN ZWEITER ORDNUNG
DISSERTATION, INTERNER BERICHT DES INST. F.
INFORMATIK 71/6, UNIVERSITAET KARLSRUHE (1971)

268* HEIDT, M.
ZUR NUMERISCHEN LOESUNG GEWOEHNLICHER
DIFFERENTIALGLEICHUNGEN ZWEITER ORDNUNG (NOTIZ)
Z. ANGEW. MATH. MECH. 53, T198 (1973)

269* HEIDT, M.
DIE NUMERISCHE LOESUNG LINEARER
DIFFERENTIALGLEICHUNGEN ZWEITER ORDNUNG MIT
KONSTANTEN KOEFFIZIENTEN
COMPUTING 13, 143-154 (1974)

270* HEIGL, F.
EINIGE SCHRANKEN FUER DIE ABSOLUTBETRAEGE DER
WURZELN ALGERAISCHER GLEICHUNGEN
MONATSHEFTE FUER MATHEMATIK 63, 287-297 (1959)

271* HEINDL, G.; REINHART, E.
A GENERAL METHOD FOR THE COMPUTATION OF
MINIMAX-ERRORS
SUBMITTED TO THE 6TH SYMPOSIUM
ON MATHEMATICAL GEODESY (3RD HOTINE SYMPOSIUM),
SIENA, APRIL 1975

272* HEINDL, G.; REINHART, E.
ADJUSTMENT BY THE PRINCIPLE OF MINIMAL MAXIMUM
ERROR
CONTRIBUTION NO.124 OF 'LEHRSTUHL F. ASTRON. U.
PHYSIK. GEODAESIE:, TECHN. UNIVERSITAET MUENCHEN

HENRICI, P.
SIEHE AUCH: GARGANTINI, I.; HENRICI, P.

273* HENRICI, P.
UNIFORMLY CONVERGENT ALGORITHMS FOR THE
SIMULTANEOUS DETERMINATION OF ALL ZEROS OF A
POLYNOMIAL
STUDIES IN NUMERICAL ANALYSIS 2, 1-8 NUMERICAL
SOLUTIONS OF NONLINEAR PROBLEMS, KONGRESSBERICHT
1968 HERAUSGEBER: ORTEGA, J.M.; RHEINBOLDT, W.C.

274* HENRICI, P.
CIRCULAR ARITHMETIC AND THE DETERMINATION OF
POLYNOMIAL ZEROS
'CONFERENCE ON APPLICATION OF NUMERICAL ANALYSIS',
LECTURE NOTES IN MATHEMATICS 228, SPRINGER VERLAG,
86 - 92 (1971)

275* HENRICI, P.
COMPUTATIONAL COMPLEX ANALYSIS
PAPER CONTRIBUTED TO THE AMS-MAA CONFERENCE ON THE
INFLUENCE OF COMPUTING ON MATHEMATICAL RESEARCH
AND EDUCATION, MISSOULA, MONTANA (1973)

HENRICI-HERZBERGER

276* HENRICI, P.
APPLIED AND COMPUTATIONAL COMPLEX ANALYSIS
WILEY-INTERSCIENCE, NEW YORK, (1974)

277* HENRICI, P.
EINIGE ANWENDUNGEN DER KREISSCHEIBENARITHMETIK IN
DER KETTENBRUCHTHEORIE
'INTERVAL MATHEMATICS', ED. BY K. NICKEL
LECTURE NOTES IN COMPUTER SCIENCE 29, SPRINGER
VERLAG, 19-30 (1975)

278* HENRICI, P.; GARGANTINI, I.
UNIFORMELY CONVERGENT ALGORITHMS FOR THE
SIMULTANEOUS APPROXIMATION OF ALL ZEROS OF A
POLYNOMIAL
'CONSTRUCTIVE ASPECTS OF THE FUNDAMENTAL THEOREM
OF ALGEBRA', ED. BY B.DEJON AND P.HENRICI
WILEY-INTERSCIENCE, LONDON, 77-113 (1969)

279* HERMES, H.
CALCULUS OF SET VALUED FUNCTIONS AND CONTROL
INDIANA UNIV. MATH. J. (J. MATH. MECH.) 18, 47-59
(1968)

280* HEROLD, G.
DAS GAUSS'SCHE ELIMINATIONSVERFAHREN ZUR LOESUNG
EINES BELIEBIGEN LINEAREN GLEICHUNGSSYSTEMS UNTER
VERWENDUNG EINER TRIPLEX-ARITHMETIK
DIPLOMARBEIT, INST. F. INFORMATIK, UNIVERSITAET
KARLSRUHE (1969)

HERZBERGER, J.
SIEHE AUCH: ALEFELD, G.; HERZBERGER, J.
ALEFELD, G.; HERZBERGER, J.;
MAYER, O.

281* HERZBERGER, J.
METRISCHE EIGENSCHAFTEN VON MENGENSYSTEMEN UND
EINIGE ANWENDUNGEN
DISSERTATION, UNIVERSITAET KARLSRUHE (1969)

282* HERZBERGER, J.
INTERVALLMAESSIGE AUSWERTUNG VON
STANDARDFUNKTIONEN IN ALGOL-60
COMPUTING 5, 377-384 (1970)

283* HERZBERGER, J.
DEFINITION UND EIGENSCHAFTEN ALLGEMEINER
INTERVALLRAEUME
Z. ANGEW. MATH. MECH. 50, T50-T51 (1970)

284* HERZBERGER, J.
ZUR KONVERGENZ INTERVALLMAESSIGER
ITERATIONSVERFAHREN
Z. ANGEW. MATH. MECH. 51, T56 (1971)

285* HERZBERGER, J.
UEBER OPTIMALE VERFAHREN ZUR NUMERISCHEN
BESTIMMUNG REELLER NULLSTELLEN MIT FEHLERSCHRANKEN
HABILITATIONSSCHRIFT, UNIVERSITAET KARLSRUHE
(1972)

286* HERZBERGER, J.
UEBER EIN VERFAHREN ZUR BESTIMMUNG REELLER
NULLSTELLEN MIT ANWENDUNG AUF PARALLELRECHNUNG
ELEKTRON. RECHENANLAGEN 14, 250-254 (1972)

287* HERZBERGER, J.
UEBER DIE NULLSTELLENBESTIMMUNG BEI
NAEHERUNGSWEISE BERECHNETEN FUNKTIONEN
COMPUTING 10, 23-31 (1972)

288* HERZBERGER, J.
BEMERKUNGEN ZU EINEM VERFAHREN VON R.E.MOORE
Z. ANGEW. MATH. MECH. 53, 356-358 (1973)

289* HERZBERGER, J.
ZUR APPROXIMATION DES WERTEBEREICHES REELLER
FUNKTIONEN DURCH INTERVALLAUSDRUECKE
COMPUTING SUPPLEMENTUM 1, ALBRECHT, R.;
KULISCH, U.; SPRINGER-VERLAG WIEN, NEW YORK,
57-64 (1977)

HITCHINS, G. D.
SIEHE AUCH: GRANT, J. A.; HITCHINS, G. D.

290* HITCHINS, G. D.
THE NUMERICAL SOLUTION OF POLYNOMIAL EQUATIONS
WITH SPECIAL REFERENCE TO THE DETERMINATION OF
INITIAL APPROXIMATIONS TO THE ROOTS
PH. D. THESIS, UNIVERSITY OF LEEDS (1971)

291* HITCHINS, G. D.
AN INTERVAL ARITHMETIC PACKAGE AND SOME
APPLICATIONS
TECHNICAL REPORT 10, CENTRE FOR COMPUTER STUDIES,
UNIVERSITY OF LEEDS (1972)

HOFFBERG, S.
SIEHE: GOLDSTEIN, M.; HOFFBERG, S.

HSU, N.-T.
SIEHE: BINSTOCK, W.; HAWKES, J.; HSU, N.-T.

292* HUBER, P.J.
KAPAZITAETEN STATT WAHRSCHEINLICHKEITEN? GEDANKEN
ZUR GRUNDLEGUNG DER STATISTIK
JBER. D. DT. MATH.-VEREIN 78, 81-92 (1976)

293* HUNGER, S.
INTERVALLANALYTISCHE DEFEKTABSCHAETZUNG BEI
ANFANGSWERTAUFGABEN FUER SYSTEME GEWOEHNLICHER
DIFFERENTIALGLEICHUNGEN
BERICHTE DER GMD BONN 41,   (1971)

294* HUNGER, S.
INTERVALLANALYTISCHE DEFEKTABSCHAETZUNG ZUR
LOESUNG MIT EXAKTER FEHLERERFASSUNG BEI
ANFANGSWERTAUFGABEN FUER SYSTEME GEWOEHNLICHER
DIFFERENTIALGLEICHUNGEN
Z. ANGEW. MATH. MECH. 52, T208-T209 (1972)

295* HUSSAIN, F.
ZUR BEDEUTUNG DER INTERVALLANALYSIS BEI
NUMERISCH-GEODAETISCHEN RECHNUNGEN
AVN 4, 128-132 (1971)

296* JACKSON, L.W.
A COMPARISON OF ELLIPSOIDAL AND INTERVAL
ARITHMETIC ERROR
BOUNDS; NUMERICAL SOLUTIONS OF NONLINEAR PROBLEMS
(NOTIZ) SIAM REV. 11, 114 (1969)

297* JACKSON, L. W.
AUTOMATIC ERROR ANALYSIS FOR THE SOLUTION OF
ORDINARY DIFFERENTIAL EQUATIONS
TECHNICAL REPORT # 28, DEPARTM. OF COMPUT.
SCIENCE, UNIVERSITY OF TORONTO (1971)

298* JACKSON, L. W.
INTERVAL ARITHMETIC ERROR-BOUNDING ALGORITHMS
SIAM J. NUMER. ANAL. 12, 223-238 (1975)

299* JACOBS, D.
THE STATE OF THE ART IN NUMERICAL ANALYSIS
PROCEEDINGS OF THE CONFERENCE ON THE STATE
OF THE ART IN NUMERICAL ANALYSIS
HELD AT THE UNIVERSITY OF YORK,
APRIL 12TH-15TH, 1976
ACADEMIC PRESS, LONDON; NEW YORK; SAN FRANCISCO
(1977)

JACOBS, M.Q.
SIEHE: BANKS, H.T.; JACOBS, M.Q.

300* JAHN, K.-U.
AUFBAU EINER 3-WERTIGEN LINEAREN ALGEBRA UND
AFFINEN GEOMETRIE AUF DER GRUNDLAGE DER
INTERVALL-ARITHMETIK
DISSERTATION, UNIVERSITAET LEIPZIG (1971)

301* JAHN, K.-U.
EINE THEORIE DER GLEICHUNGSSYSTEME MIT
INTERVALLKOEFFIZIENTEN
Z. ANGEW. MATH. MECH. 54, 405-412 (1974)

JAHN–KAHAN

302* JAHN, K.-U.
INTERVALL-WERTIGE MENGEN
MATH. NACHR. 65, 105-116 (1975)

303* JAHN, K.-U.
PUNKTKONVERGENZ IN DER INTERVALL-RECHNUNG
Z. ANGEW. MATH. MECH. 55, 606-608 (1975)

304* JAHN, K.-U.
EINE AUF DEN INTERVALLZAHLEN FUSSENDE 3-WERTIGE
LINEARE ALGEBRA
MATH. NACHRICHTEN 65, 105-116 (1975)

305* JAHN, K.U.
MIN - MAX - INTERVALLRECHNUNG
MATH. NACHRICHTEN 71, 267-272 (1976)

306* JAHN, K.-U.
DIE INTERVALL-ARITHMETIK ALS BASIS EINER
MEHRWERTIGEN ANALYSIS
DISSERTATION ZUR PROMOTION B ANGENOMMEN AUF
BESCHLUSS DES SENATS DER KARL-MARX-UNIVERSITAET
LEIPZIG (1976)

JASCHKE, H.R.
SIEHE: MOORE, R.E.; DAVIDSON, J.A.;
JASCHKE, H.R.; SHAYER, S.

307* JUELLIG, W.
NUMERISCHE BERECHNUNG VON SCHRANKEN FUER DEN
WERTEBEREICH EINES TRIGONOMETRISCHEN POLYNOMS IN
EINEM INTERVALL
DIPLOMARBEIT, INST. F. INFORMATIK, UNIVERSITAET
KARLSRUHE (1972)

JULDASEV, Z.CH.
SIEHE: SHOKIN, Y.; JULDASEV, Z. CH.

308* KAHAN, W. M.
A COMPUTABLE ERROR-BOUND FOR SYSTEMS OF ORDINARY
DIFFERENTIAL EQUATIONS (ABSTRACT)
SIAM REV. 8, 568-569 (1966)

309* KAHAN, W. M.
CIRCUMSCRIBING AN ELLIPSOID ABOUT THE MINKOWSKI
SUM OF GIVEN ELLIPSOIDS
ERSCHEINT IN J. LINEARE ALGEBRA (1967)

310* KAHAN, W. M.
AN ELLIPSOIDAL ERROR-BOUND FOR LINEAR SYSTEMS OF
ORDINARY DIFFERENTIAL EQUATIONS
UNVEROEFFENTLICHTES MANUSKRIPT (1967)

311* KAHAN, W. M.
CIRCUMSCRIBING AN ELLIPSOID ABOUT THE INTERSECTION
OF TWO ELLIPSOIDS
CAN. MATH. BULL. 11, 437-441 (1968)

312* KAHAN, W. M.
AN ELLIPSOIDAL ERROR BOUND FOR LINEAR SYSTEMS OF
ORDINARY DIFFERENTIAL EQUATIONS
ERSCHEINT (1968)

313* KAHAN, W.M.
A MORE COMPLETE INTERVAL ARITHMETIC LECTURE NOTES
FOR AN ENGINEERING SUMMER COURSE IN
NUMERICAL ANALYSIS AT THE UNIVERSITY OF MICHIGAN,
UNIVERSITY OF MICHIGAN (1968)

314* KAHAN, W.M.
INVITED COMMENTARY (CONCERNING THE INVITED PAPER
OF NICKEL, K.: ERROR BOUNDS AND COMPUTER
ARITHMETIC)
"PROC. OF IFIP CONGRESS 1968, VOL. I", ED. BY A.
MORRELL NORTH-HOLLAND PUBL. COMP., AMSTERDAM,
60-62 (1969)

315* KAHAN, W.M.
A SURVEY OF ERROR ANALYSIS
"PROC. OF IFIP CONGRESS 1971, VOL. II" ED. BY
C.V.FREIMAN, J.E.GRIFFITH AND J.L.ROSENFELD
NORTH-HOLLAND PUBL. COMP., AMSTERDAM, 1214-1239
(1972)

KAHAN-KAUCHER

316* KAHAN, W. M.
ELLIPSOIDAL BOUNDS FOR THE PROPAGATION OF
UNCERTAINTY ALONG TRAJECTORIES
PRESENTATION AT THE CONFERENCE
ON THE NUMERICAL SOLUTION OF
ORDINARY DIFFERENTIAL EQUATIONS,
UNIVERSITY OF TEXAS, AUSTIN (1972)

317* KAHAN, W. M.
IMPLEMENTATION OF ALGORITHMS (CHAPTER 12)
REPORT AD 769-124, NATIONAL TECHNICAL INFORMATION
SERVICE, SPRINGFIELD, VIRGINIA

KAHLERT-WARMBOLD, I.
SIEHE AUCH: NUDING, E.; WARMBOLD, I.

318* KAHLERT-WARMBOLD, I.
SIMULATIVE UND KONSTRUKTIVE PROBLEME
Z. ANGEW. MATH. MECH. 56, T293-T295 (1976)

319* KALDENBACH, P.
LINEARE GLEICHUNGSSYSTEME UND TRIPLEX-ALGOL
DIPLOMARBEIT AM INSTITUT FUER INFORMATIK,
UNIVERSITAET KARLSRUHE (1969)

320* KANNER, H.
NUMBER BASE CONVERSION IN A SIGNIFICANT DIGIT
ARITHMETIC
J. ASSOC. COMPUT. MACH. 12, 242-246 (1965)

321* KANSY, K.
EIN INTERVALLARITHMETISCHES VERFAHREN ZUR
BESTIMMUNG ALLER NULLSTELLEN EINER FUNKTION IN
EINEM REELLEN INTERVALL
DIPLOMARBEIT, UNIVERSITAET BONN (1969)

322* KANSY, K.
ABLEITUNGSVERTRAEGLICHE VERALLGEMEINERUNG DER
INTERVALLPOLYNOME
BERICHTE DER GMD BONN 70 (1973)

323* KANSY, K.
INTERVALLARITHMETISCHE DARSTELLUNG VON FUNKTIONEN
UND IHREN ABLEITUNGEN
Z. ANGEW. MATH. MECH. 54, T222 (1974)

324* KANSY, K.
IMPLEMENTIERUNG DER INTERVALLARITHMETIK AUF DEN
RECHENANLAGEN
IBM/360, IBM/370 UND SIEMENS 4C04 ERSCHEINT IN
MITTEILUNGEN DER GMD, BONN

325* KANSY, K.; KRUECKEBERG, F.; SCHUETH, D.;
SCHUHMACHER, H.; TOFAHRN, W.
BESCHREIBUNG DES POLYNOMBAUKASTENS
PROGRAMMBIBLIOTHEK DER GMD BONN

326* KARTHEUS, V.
INNENINTERVALL-DARSTELLUNG DER LOESUNGSMENGE
LINEARER INTERVALL-GLEICHUNGSSYSTEME
DIPLOMARBEIT AM INST. F. ANGEW. MATH. U.
INFORMATIK, UNIVERSITAET BONN (1970)

327* KARTHEUS, V.
ZUR INTERVALLANALYTISCHEN BEHANDLUNG LINEARER
GLEICHUNGSSYSTEME
MITTEILUNGEN DER GMD BONN 16 (1972)

328* KATZAN, P.
FAKTORZERLEGUNG VON POLYNOMEN MIT FEHLERERFASSUNG
BERICHTE DER GMD BONN 18 (1969)

329* KAUCHER, E.
UEBER METRISCHE UND ALGEBRAISCHE EIGENSCHAFTEN
EINIGER BEIM NUMERISCHEN RECHNEN AUFTRETENDER
RAEUME
DISSERTATION, UNIVERSITAET KARLSRUHE (1973)

330* KAUCHER, E.
ALLGEMEINE EINBETTUNGSSAETZE ALGEBRAISCHER
STRUKTUREN UNTER ERHALTUNG VON VERTRAEGLICHEN
ORDNUNGS- UND VERBANDSSTRUKTUREN MIT ANWENDUNG DER
INTERVALLRECHNUNG
Z. ANGEW. MATH. MECH. 56, T296-T297 (1976)

331* KAUCHER, E.
ALGEBRAISCHE ERWEITERUNGEN DER INTERVALLRECHNUNG
UNTER ERHALTUNG DER OPDNUNGS- UND
VERBANDSSTRUKTUREN
COMPUTING SUPPLEMENTUM 1, ALBRECHT, R.;
KULISCH, U., SPRINGER-VERLAG WIEN, NEW YORK,
65-79 (1977)

332* KAUCHER, E.
UEBER EIGENSCHAFTEN UND ANWENDUNGSMOEGLICHKEITEN
DER ERWEITERTEN INTERVALLRECHNUNG UND DES
HYPERBOLISCHEN FASTKOERPERS UEBER R
COMPUTING SUPPLEMENTUM 1, ALBRECHT, R.;
KULISCH, U., SPRINGER-VERLAG WIEN, NEW YORK,
81-94 (1977)

KELLER, C. L.
SIEHE AUCH: BREITER, M.C.; KELLER, C. L.;
REEVES, T. E.
GUDERLEY, K. G.; KELLER, C. L.

333* KELLER, C. L.; REEVES, T. E.
COMPUTING ERROR BOUNDS WHEN THE TRUNCATION ERROR
AND PROPAGATED ERROR ARE OF THE SAME ORDER
REPORT ARL 70-C066 AEROSPACE RESEARCH LABS.,
WRIGHT-PATTERSON AIR FORCE BASE, OHIO (1970)

334* KIMBLE, G. W.
COMPUTATIONAL ERROR
'INFORMATION AND COMPUTER SCIENCE' ED. BY
G.W.KIMBLE HOLT, RINEHART AND WINSTON, INC. NEW
YORK (1975)

335* KLATTE, R.
ANWENDUNGEN EINER VERALLGEMEINERUNG DER
MODIFIKATION DES NEWTONVERFAHRENS
DIPLOMARBEIT AM INST. F. ANGEW. MATH.,
UNIVERSITAET KARLSRUHE (1969)

336* KLAUA, D.
PARTIELL DEFINIERTE MENGEN
MBER. DT. AKAD. WISS. 10, 571-578 (1968)

337* KLAUA, D.
PARTIELLE MENGEN MIT MEHRWERTIGEN GRUNDPEZIEHUNGEN
MBER. DT. AKAD. WISS. 11, 573-584 (1969)

338* KLAUA, D.
PARTIELLE MENGEN UND ZAHLEN
MBER. DT. AKAD. WISS. 11, 585-599 (1969)

339* KLAUA, D.
INTERVALLSTRUKTUREN GEORDNETER KOERPER
MATH. NACHRICHTEN 75, 319-326 (1976)

340* KLAUA, D.
PARTIELLE BOOLESCHE ALGEBREN IN DER
INTERVALLMATHEMATIK
EINGEREICHT BEI MATH. NACHR.

341* KLAUA, D.
PARTIELLE BOOLESCHE ALGEBREN IN DER
INTERVALLMATHEMATIK II
MATH. NACHRICHTEN
ZUM DRUCK ANGENOMMEN

342* KLEIN, H.-O.
INVERSION VON PARAMETERABHAENGIGEN MATRIZEN
DIPLOMARBEIT AM INST. F. ANGEW. MATH. U.
INFORMATIK, UNIVERSITAET BONN (1968)

343* KLEIN, H.-O.
EINE INTERVALLANALYTISCHE METHODE ZUR LOESUNG VON
INTEGRALGLEICHUNGEN MIT MEHRPARAMETRIGEN
FUNKTIONEN
BERICHTE DER GMD BONN 69 (1973)

344* KNOETTNER, H.
RECHNEN MIT EINER KLASSE VON FUNKTIONSINTERVALLEN
DIPLOMARBEIT AM INST. F. ANGEW. MATHEMATIK
UNIVERSITAET HEIDELBERG (1977)

KOPP-KRAWCZYK

345* KOPP, G.
DIE NUMERISCHE BEHANDLUNG VON REELLEN LINEAREN
GLEICHUNGSSYSTEMEN MIT FEHLERERFASSUNG FUER
M-MATRIZEN SOWIE FUER DIAGONALDOMINANTE UND
INVERS-ISOTONE MATRIZEN
DIPLOMARBEIT, ANGEFERTIGT AM INST. F. PRAKT.
MATH., UNIV. KARLSRUHE (1976)

346* KRACHT, M.; SCHROEDER, G.
ZUR INTERVALLRECHNUNG IN LINEAREN RAEUMEN
COMPUTING 11, 73-79 (1973)

347* KRACHT, M.; SCHROEDER, G.
EINE EINFUEHRUNG IN DIE THEORIE DER QUASILINEAREN
RAEUME MIT ANWENDUNG AUF DIE IN DER
INTERVALLRECHNUNG AUFTRETENDEN RAEUME
MATH.-PHYS. SEMESTERBER. XX, 226-242 (1973)

348* KRAFT, M.
INTERVALLARITHMETISCHE RELAXATIONSVERFAHREN ZUM
LOESEN VON LINEAREN GLEICHUNGSSYSTEMEN
DIPLOMARBEIT AM RECHENZENTRUM, UNIVERSITAET
HEIDELBERG (1973)

KRAWCZYK, R.
SIEHE AUCH: WIPPERMANN, H.-W. ET AL.

349* KRAWCZYK, R.
FEHLERABSCHAETZUNG REELLER EIGENWERTE UND
EIGENVEKTOREN VON MATRIZEN
INTERNER BERICHT DES INST. F. INFORMATIK 68/5,
UNIVERSITAET KARLSRUHE (1968)
PUBLIZIERT IN:
COMPUTING 4, 281-293 (1969)

350* KRAWCZYK, R.
NEWTON-ALGORITHMEN ZUR BESTIMMUNG VON NULLSTELLEN
MIT FEHLERSCHRANKEN
INTERNER BERICHT DES INST. F. INFORMATIK 68/6,
UNIVERSITAET KARLSRUHE (1968)
PUBLIZIERT IN:
COMPUTING 4, 187-201 (1969)

351* KRAWCZYK, R.
ITERATIVE VERBESSERUNG VON SCHRANKEN FUER
EIGENWERTE UND EIGENVEKTOREN REELLER MATRIZEN
INTERNER BERICHT DES INST. F. INFORMATIK 68/8,
UNIVERSITAET KARLSRUHE (1968)
PUBLIZIERT IN:
Z. ANGEW. MATH. MECH. 48, T80-T83 (1968)

352* KRAWCZYK, R.
ITERATIONSVERFAHREN ZUR BESTIMMUNG KOMPLEXER
NULLSTELLEN
INTERNER BERICHT DES INST. F. INFORMATIK 69/4,
UNIVERSITAET KARLSRUHE (1969)
PUBLIZIERT IN:
Z. ANGEW. MATH. MECH. 50, T58-T61 (1970)

353* KRAWCZYK, R.
VERBESSERUNG VON SCHRANKEN FUER EIGENWERTE UND
EIGENVEKTOREN VON MATRIZEN
INTERNER BERICHT DES INST. F. INFORMATIK 69/5,
UNIVERSITAET KARLSRUHE (1969)
PUBLIZIERT IN:
COMPUTING 5, 200-206 (1970)

354* KRAWCZYK, R.
APPROXIMATION DURCH INTERVALLFUNKTIONEN
INTERNER BERICHT DES INST. F. INFORMATIK 69/7,
UNIVERSITAET KARLSRUHE (1969)

355* KRAWCZYK, R.
EINSCHLIESSUNG VON NULLSTELLEN MIT HILFE EINER
INTERVALLARITHMETIK
COMPUTING 5, 356-370 (1970)

356* KRAWCZYK, R.
ITERATIONSVERFAHREN IN HALBGEORDNETEN RAEUMEN
INTERNER BERICHT DES INST. F. INFORMATIK 70/2,
UNIVERSITAET KARLSRUHE (1970)

KRAWCZYK-KRIER

357* KRAWCZYK, R.
GLEICHUNGEN IN HALBGEORDNETEN RAEUMEN
INTERNER BERICHT DES INST. F. INFORMATIK 70/3,
UNIVERSITAET KARLSRUHE (1970)
PUBLIZIERT IN:
ABH. MATH. SEM. UNIV. HAMBURG 36,
150-165 (1971)

358* KRAWCZYK, R.
ZUR EINSCHLIESSUNG INVERSER LINEARER OPERATOREN
INTERNER BERICHT DES INST. F. INFORMATIK 70/18,
UNIVERSITAET KARLSRUHE (1970)
PUBLIZIERT IN:
'METHODEN UND VERFAHREN DER MATH. PHYS. 7'
HERAUSG. B. BROSOWSKI, E. MARTENSEN
BIBL. INST. MANNHEIM, 1-19 (1972)

359* KRAWCZYK, R.
ITERATIONSVERFAHREN IN HALBGEORDNETEN RAEUMEN
(NOTIZ)
Z. ANGEW. MATH. MECH. 51, T62 (1971)

360* KRAWCZYK, R.
FEHLERABSCHAETZUNG BEI NICHTLINEAREN
GLEICHUNGSSYSTEMEN
INTERNER BERICHT DES INST. F. PRAKT. MATH. 73/3,
UNIVERSITAET KARLSRUHE (1973)
PUBLIZIERT IN:
COMPUTING 11, 365-377 (1973)

361* KRAWCZYK, R.
FEHLERABSCHAETZUNG BEI LINEARER OPTIMIERUNG
'INTERVAL MATHEMATICS', ED. BY K. NICKEL
LECTURE NOTES IN COMPUTER SCIENCE 29, SPRINGER
VERLAG, 215-222 (1975)

362* KREIFELTS, T.
UEBER VOLLAUTOMATISCHE ERFASSUNG UND ABSCHAETZUNG
VON RUNDUNGSFEHLERN IN ARITHMETISCHEN PROZESSEN
BERICHTE DER GMD BONN 62 (1972)

363* KRESS, O.
NUMERISCHE BERECHNUNG DES DOMINANTEN EIGENWERTES

EINER MATRIX MIT FEHLERSCHRANKEN
DIPLOMARBEIT AM INST. F. INFORMATIK, UNIVERSITAET
KARLSRUHE (1969)

364* KRESS, O.
EINSCHRANKUNG VON EIGENWERTEN REELLSYMMETRISCHER
MATRIZEN
INTERNER BERICHT DES INST. F. INFORMATIK 70/10,
UNIVERSITAET KARLSRUHE (1970)

365* KRESS, O.
UEBER EIN VERFAHREN ZUR ITERATIVEN VERBESSERUNG
VON EIGENVEKTOR- UND EIGENWERTSCHRANKEN
INTERNER BERICHT DES INST. F. INFORMATIK 72/2,
UNIVERSITAET KARLSRUHE (1972)

366* KRESS, O.
UEBER FEHLERABSCHAETZUNGEN FUER EIGENWERTE UND
EIGEN-VEKTOREN REELLSYMMETRISCHER MATRIZEN
DISSERTATION, INTERNER BERICHT DES INST. F. PRAKT.
MATH. 73/4, UNIVERSITAET KARLSRUHE (1973)

KRIER, N.
SIEHE AUCH: SPELLUCCI, P.; KRIER, N.

367* KRIER, N.
KOMPLEXE KREISARITHMETIK
DISSERTATION, INTERNER BERICHT DES INST. F. PRAKT.
MATH. 73/2, UNIVERSITAET KARLSRUHE (1973)

368* KRIER, N.
KOMPLEXE KREISARITHMETIK
Z. ANGEW. MATH. MECH. 54, T225-T226 (1974)

369* KRIER, N.; SPELLUCCI, P.
EINSCHLIESSUNGSMENGEN VON POLYNOM-NULLSTELLEN
'INTERVAL MATHEMATICS', ED. BY K. NICKEL
LECTURE NOTES IN COMPUTER SCIENCE 29, SPRINGER
VERLAG, 223-228 (1975)

KRIER-KRUECKEBERG

370* KRIER, N.; SPELLUCCI, P.
EIN VERFAHREN ZUR BEHANDLUNG VON
AUSGLEICHSAUFGABEN MIT INTERVALLKOEFFIZIENTEN
PREPRINT, ERSCHEINT IN COMPUTING

KRUECKEBERG, F.
SIEHE AUCH: KANSY, K. ET AL.

371* KRUECKEBERG, F.
ZUR NUMERISCHEN INTEGRATION UND FEHLERERFASSUNG
BEI ANFANGSWERTAUFGABEN GEWOEHNLICHER
DIFFERENTIALGLEICHUNGEN
DISSERTATION, SCHRIFTEN D. RH.-W. INST. F.
INSTRUM. MATH. 1, UNIVERSITAET BONN (1961)

372* KRUECKEBERG, F.
ZUR NUMERISCHEN INTEGRATION UND FEHLERRECHNUNG BEI
GEWOEHNLICHEN DIFFERENTIALGLEICHUNGEN
'MATHEMATISCHE METHODEN D. HIMMELSMECHANIK U.
ASTRONAUTIK' HERAUSG. E. STIEFEL,
BIBL.INST.MANNHEIM (1966)

373* KRUECKEBERG, F.
NUMERISCHE INTERVALLRECHNUNG UND DEREN ANWENDUNG
INTERNER BERICHT D. RH.-W. INST. F. INSTRUM.
MATH., UNIVERSITAET BONN (1966)

374* KRUECKEBERG, F.
INVERSION VON MATRIZEN MIT FEHLERERFASSUNG
Z. ANGEW. MATH. MECH. 46, 169-171 (1966)

375* KRUECKEBERG, F.
ZUR NUMERISCHEN LOESUNG VON INTEGRALGLEICHUNGEN
MIT ELEKTRONISCHEN DATENVERARBEITUNGSANLAGEN
SITZUNGSBERICHTE DER BERLINER MATH. GES. 52
(1967-1968)

376* KRUECKEBERG, F.
DEFEKTERFASSUNG BEI GEWOEHNLICHEN UND PARTIELLEN
DIFFERENTIALGLEICHUNGEN
NUMERISCHE MATHEMATIK, DIFFERENTIALGLEICHUNGEN,
APPROXIMATIONSTHEORIE', ISNM 19, HERAUS.
L.COLLATZ, G. MEINARDUS, H. UNGER BIRKHAEUSER
VERLAG, STUTTGART, BASEL, 69-82 (1968)

377* KRUECKEBERG, F.
BEMERKUNGEN ZUR INTERVALL-ANALYSIS
APL. MAT. 13, 152-153 (1968)

378* KRUECKEBERG, F.
ORDINARY DIFFERENTIAL EQUATIONS
'TOPICS IN INTERVAL ANALYSIS', ED. BY E. HANSEN,
OXFORD UNIVERSITY PRESS, 91-97 (1969)

379* KRUECKEBERG, F.
PARTIAL DIFFERENTIAL EQUATIONS
'TOPICS IN INTERVAL ANALYSIS', ED. BY E. HANSEN,
OXFORD UNIVERSITY PRESS, 98-101 (1969)

380* KRUECKEBERG, F.
INTERVALLANALYTISCHE METHODEN IN DER NUMERISCHEN
DATENVERARBEITUNG
JAHRBUCH 1970, WESTDTSCH. VERLAG OPLADEN, 233-261
(1970)

381* KRUECKEBERG, F.
IMPLEMENTATION AND FORMALIZATION OF FLOATING-POINT
ARITHMETIC
CARATHEODORY SYMPOSIUM, ATHEN (1973)

382* KRUECKEBERG, F.; UNGER, H.
ON THE NUMERICAL INTEGRATION OF ORDINARY
DIFFERENTIAL EQUATIONS AND THE DETERMINATION OF
ERROR BOUNDS
'SYMPOSIUM ON THE NUMERICAL TREATMENT OF ORDINARY
DIFFERENTIAL EQUATIONS, INTEGRAL AND
INTEGRO-DIFFERENTIAL EQUATIONS, PROC. OF THE ROME
SYMPOSIUM (20-24 SEPT 1960)' ED. BY PROVISIONAL
INTERNAT. COMPUTATION CENTRE BIRKHAEUSER VERLAG,
BASEL, STUTTGART, 369-379 (1960)

KRULLIS-KULISCH

383* KRULLIS, S.
PROGRAMMIERUNG EINER VERKETTUNGSMATRIX VON
LITERATURZITATEN IN ALGOL 60
SEMESTERARBEIT AM INST. F. INFORMATIK,
UNIVERSITAET STUTTGART (1972)

384* KUBA, D.; RALL, L.B.
A UNIVAC 1108 PROGRAM FOR OBTAINING RIGOROUS ERROR
ESTIMATES FOR APPROXIMATE SOLUTIONS OF SYSTEMS OF
EQUATIONS
MRC TECHNICAL SUMMARY REPORT # 1168 UNIVERSITY OF
WISCONSIN, MADISON (1972)

KUBO, T.
SIEHE AUCH: OHNAKA, K.; YASUI, H.; KUBO, T.

KULISCH, U.
SIEHE AUCH: ALBRECHT, R.; KULISCH, U.
APOSTOLATOS, N.; KULISCH, U.
APOSTOLATOS, N.; KULISCH, U.;
NICKEL, K.
WIPPERMANN, H.-W. ET AL.

385* KULISCH, U.
GRUNDLAGEN EINER MASCHINENINTERVALLARITHMETIK
(NOTIZ)
Z. ANGEW. MATH. MECH. 47, T81 (1967)

386* KULISCH, U.
GRUNDZUEGE DER INTERVALLRECHNUNG (NOTIZ)
Z. ANGEW. MATH. MECH. 48, T111 (1968)

387* KULISCH, U.
GRUNDZUEGE DER INTERVALLRECHNUNG
"UEBERBLICKE MATHEMATIK 2", HERAUSG. L. LAUGWITZ
BIBL. INST. MANNHEIM, ZUERICH, 51-98 (1969)

388* KULISCH, U.
AN AXIOMATIC APPROACH TO ROUNDED COMPUTATIONS
MRC TECHNICAL SUMMARY REPORT # 1020 UNIVERSITY OF
WISCONSIN, MADISON (1969)
PUBLIZIERT IN:
NUMER. MATH. 18, 1-17 (1971)

389* KULISCH, U.
ON THE CONCEPT OF A SCREEN
MRC TECHNICAL SUMMARY REPORT # 1084 UNIVERSITY OF
WISCONSIN, MADISON (1970)
PUBLIZIERT IN:
Z. ANGEW. MATH. MECH. 53, 115-119 (1973)

390* KULISCH, U.
ROUNDING INVARIANT STRUCTURES
MRC TECHNICAL SUMMARY REPORT # 1103 UNIVERSITY OF
WISCONSIN, MADISON (1971)

391* KULISCH, U.
INTERVAL ARITHMETIC OVER COMPLETELY ORDERED
RINGOIDS
MRC TECHNICAL SUMMARY REPORT # 1105 UNIVERSITY OF
WISCONSIN, MADISON (1972)

392* KULISCH, U.
AXIOMATIK DES NUMERISCHEN RECHNENS (NOTIZ)
Z. ANGEW. MATH. MECH. 52, T211 (1972)

393* KULISCH, U.
IMPLEMENTATION AND FORMALIZATION OF FLOATING-POINT
ARITHMETIC
CARATHEODORY SYMPOSIUM, ATHEN (1973)

394* KULISCH, U.
UEBER DIE ARITHMETIK VON RECHENANLAGEN
JAHRBUCH UEBERBLICKE MATHEMATIK 1975, 69-108
(1975)

395* KULISCH, U.
FORMALIZATION AND IMPLEMENTATION OF FLOATING-POINT
ARITHMETICS
COMPUTING 14, 323-348 (1975)

396* KULISCH, U.
GRUNDLAGEN DES NUMERISCHEN RECHNENS MATHEMATISCHE
BEGRUENDUNG DER RECHNERARITHMETIK
REIHE INFORMATIK, 19 (1976) BIBLIOGRAPHISCHES
INSTITUT, MANNHEIM

397* KULISCH, U.
EIN KONZEPT FUER EINE ALLGEMEINE THEORIE DER
RECHNERARITHMETIK
COMPUTING SUPPLEMENTUM 1, ALBRECHT, R.;
KULISCH, U., SPRINGER-VERLAG WIEN, NEW YORK,
95-105 (1977)

398* KULISCH, U.
UEBER DIE BEIM NUMERISCHEN RECHNEN MIT
RECHENANLAGEN AUFTRETENDEN RAEUME
COMPUTING SUPPLEMENTUM 1, ALBRECHT, R.;
KULISCH, U., SPRINGER-VERLAG WIEN, NEW YORK,
107-119 (1977)

399* KULISCH, U.; BOHLENDER, G.
FORMALIZATION AND IMPLEMENTATION OF FLOATING-POINT
MATRIX OPERATIONS
COMPUTING 16, 239-261 (1976)

400* KUNATH, P.
EINSCHLIESSUNG ELEMENTARER TRANSZENDENTER
FUNKTIONEN DURCH RATIONALE FUNKTIONSSTREIFEN
DIPLOMARBEIT AM INST. F. ANGEW. MATH. U.
INFORMATIK, UNIVERSITAET BONN (1969)

401* KUPERMAN, I. B.
APPROXIMATE LINEAR ALGEBRAIC EQUATIONS AND
ROUNDING ERROR ESTIMATION
PH. D. THESIS, DEPT. OF APPL. MATH., UNIVERSITY OF
WITWATERSRAND, JOHANNESBURG (1967)

402* KUPERMAN, I. B.
APPROXIMATE LINEAR ALGEBRAIC EQUATIONS
VAN NOSTRAND REINHOLD CO., LONDON, 2, (1971)

403* KUPKA, I.
SIMULATION REELLER ARITHMETIK UND REELLER
FUNKTIONEN IN ENDLICHEN MENGEN
NUMER. MATH. 17, 143-152 (1971)

404* KWAPISZ, M.
ON APPROXIMATE ITERATIONS AND SOLUTIONS FOR
EQUATIONS CONSIDERED IN FUNCTION SPACES

405* KWAPISZ, M.; TURO, J.
APPROXIMATE ITERATIONS FOR A HYPERBOLIC
FUNCTIONAL-DIFFERENTIAL EQUATION IN BANACH SPACE
DIFFERENTIAL EQUATIONS 11, 1212-1223 (1976)

LADNER, T.D.
SIEHE AUCH: CRARY, F.D.; LADNER, T.D.

406* LADNER, T. D.; YOHE, J. M.
AN INTERVAL ARITHMETIC PACKAGE FOR THE UNIVAC 1108
MRC TECHNICAL SUMMARY REPORT # 1005 UNIVERSITY OF
WISCONSIN, MADISON (1970)

407* LAI, S.
IMPLEMENTATION OF BASIC SOFTWARE FOR SIGNIFICANT
DIGIT ARITHMETIC
DEPARTMENT OF COMPUTER SCIENCE REPORT 530,
UNIVERSITY OF ILLINOIS, URBANA (1972)

408* LAIER, G.
REALISIERUNG EINER FEHLERSCHRANKEN-ARITHMETIK FUER
DIE RECHENANLAGE ZUSE Z 23
DIPLOMARBEIT AM INST. F. INFORMATIK, UNIVERSITAET
KARLSRUHE (1968)

LANCASTER, P.
SIEHE: ROKNE, J.; LANCASTER, P.

409* LATA, M.; MITTAL, B.S.
A DECOMPOSITION METHOD FOR INTERVAL LINEAR
FRACTIONAL PROGRAMMING
Z. ANGEW. MATH. MECH. 56, 153-159 (1976)

410* LAUBE, B.
NUMERISCH KONVERGENTE VERFAHREN DER NUMERISCHEN
DIFFERENTIATION
DIPLOMARBEIT AM INST. F. PRAKT. MATH.,
UNIVERSITAET KARLSRUHE (1973)

'INTERVAL MATHEMATICS', ED. BY K. NICKEL
LECTURE NOTES IN COMPUTER SCIENCE 29, SPRINGER
VERLAG, 229-235 (1975)

411* LAVEUVE, S. E.
      DEFINITION EINER KAHAN-ARITHMETIK UND IHRE
      IMPLEMENTIERUNG
      'INTERVAL MATHEMATICS', ED. BY K. NICKEL
      LECTURE NOTES IN COMPUTER SCIENCE 29, SPRINGER
      VERLAG, 236-245 (1975)

412* LAVEUVE, S. E.
      DEFINITION EINER KAHAN-ARITHMETIK UND IHRE
      IMPLEMENTIERUNG IN TRIPLEX-ALGOL 60
      DIPLOMARBEIT, INT. BER. D. INST. F. PRAKT. MATH.
      75/6, UNIVERSITAET KARLSRUHE (1975)

413* LEE, Y.D.; MOORE, R.E.
      RIGOROUS GLOBAL ERROR BOUNDS FOR NONLINEAR
      TWO-POINT BOUNDARY VALUE PROBLEMS
      (SUBMITTED TO SIAM J. NUMER. ANAL.), (1976)

414* LEKKERKERKER, G. G.; BOLAND, J.C.
      REPRESENTATION OF A FINITE GRAPH BY A SET OF
      INTERVALS ON THE REAL LINE
      FUNDAMENTA MATHEMATICAE 51, 45-64 (1962)

      LEONHARDT, M.
      SIEHE: SCHMIDT, J.W.; LEONHARDT, M.

415* LEVEN, W.
      EIN PROGRAMMSYSTEM FUER INTERVALLRECHNUNG MIT
      INNENRUNDUNG AUF DER IBM 7090 UND ANWENDUNGEN BEI
      UNTERSUCHUNGEN UEBER DEN GAUSS'SCHEN ALGORITHMUS
      UND DAS CHOLESKY-VERFAHREN IN INTERVALLARITHMETIK
      DIPLOMARBEIT AM INST. F. ANGEW. MATH. U.
      INFORMATIK, UNIVERSITAET BONN (1970)

416* LOJKO, M.S.
      COMPUTER IMPLEMENTATION OF INTERVAL ARITHMETIC AND
      ITS USE IN GRAM-SCHMIDT ORTHOGONALIZATION
      M. SC. THESIS, UNIVERSITY OF COLORADO (1960)

      LONDON, R.L.
      SIEHE: GOOD, D. I.; LONDON, R. L.

417* LORD, A.
      B-5500 ALGOL PROCEDURES FOR RANGE ARITHMETIC
      COMPUTER SCIENCE DEPARTMENT, TECHNICAL REPORT CS
      239, STANFORD UNIVERSITY (1964)

      LORTZ, B.
      SIEHE AUCH: WIPPERMANN, H.-W. ET AL.

418* LORTZ, B.
      EINE LANGZAHLARITHMETIK MIT OPTIMALER EINSEITIGER
      RUNDUNG
      DISSERTATION, UNIVERSITAET KARLSRUHE (1971)

      LOSCALZO, F.R.
      SIEHE: DARGEL, R.H.; LOSCALZO, F.R.; WITT, T.H.

419* MACHOST, B.
      NUMERISCHE BEHANDLUNG DES SIMPLEXVERFAHRENS MIT
      INTERVALLANALYTISCHEN METHODEN
      BERICHTE DER GMD BONN 30 (1970)

420* MACHOST, B.
      EINE NUMERISCH GESICHERTE BEHANDLUNG DES
      SIMPLEXVERFAHRENS MIT INTERVALLANALYTISCHEN
      METHODEN
      Z. ANGEW. MATH. MECH. 51, T64-T66 (1971)

      MADSEN, K.
      SIEHE AUCH: CAPRANI, O.; MADSEN, K.

421* MADSEN, K.
      ON THE SOLUTION OF NONLINEAR EQUATIONS IN INTERVAL
      ARITHMETIC
      BIT 13, 428-433 (1973)

422* MAIER, H.
      ZUR THEORIE UND ANWENDUNG VON NAEHERUNGSVERFAHREN
      FUER DIE LOESUNG NICHTLINEARER OPERATORGLEICHUNGEN
      DISSERTATION, UNIVERSITAET MUENCHEN (1972)

423* MAIER, M.
EMPFINDLICHKEITSANALYSE FUER DIE NULLSTELLEN EINES
KOMPLEXEN POLYNOMS
DIPLOMARBEIT AM INSTITUT FUER INFORMATIK,
UNIVERSITAET KARLSRUHE (1969)

424* MAJUMDER, K. L.; BHATTACHARJEE, G. P.
INTERVAL ARITHMETIC PACKAGE FOR IBM-1620
MANUSKRIPT (1973)

425* MAJUMDER, K.L.; BHATTACHARJEE, G. P.
SOME ALGORITHMS FOR INTERVAL INTERPOLATING
POLYNOMIAL
MANUSKRIPT (1973)
PUBLIZIERT IN: COMPUTING 16, 305-317 (1976)

MALLORY, F. B.
SIEHE: BRIGHT, H. S.; COLHOUN, B. A.;
MALLORY, F. B.

426* MANCINI, L.J.
APPLICATIONS OF INTERVAL ARITHMETIC IN SIGNOMIAL
PROGRAMMING
TECHNICAL REP. SOL 75-23, SYSTEMS OPTIMIZATION
LAB. DEPT. OF OPERATIONS RESEARCH, STANFORD
UNIVERSITY (1975)

427* MANCINI, L.J.; MC CORMICK, G.P.
BOUNDING GLOBAL MINIMA
MATH. OF OPERATIONS RESEARCH 1, 50-53 (1976)

428* MARCOWITZ, U.
FEHLERABSCHAETZUNG BEI ANFANGSWERTAUFGABEN FUER
SYSTEME VON GEWOEHNLICHEN DIFFERENTIALGLEICHUNGEN
MIT ANWENDUNG AUF DAS PROBLEM DES WIEDEREINTRITTS
EINES RAUMFAHRZEUGS IN DIE LUFTHUELLE DER ERDE
DISSERTATION, UNIVERSITAET KOELN (1973)

429* MARCOWITZ, U.
FEHLERABSCHAETZUNG BEI ANFANGSWERTAUFGABEN FUER
SYSTEME VON GEWOEHNLICHEN DIFFERENTIALGLEICHUNGEN
MIT ANWENDUNG AUF DAS REENTRY-PROBLEM
NUMER. MATH. 24, 249-275 (1975)

430* MARCOWITZ, U.
FEHLERABSCHAETZUNG BEI ANFANGSWERTAUFGABEN FUER
SYSTEME VON GEWOEHNLICHEN DIFFERENTIALGLEICHUNGEN
MIT ANWENDUNG AUF DAS REENTRY-PROBLEM
INST. F. ANGEW. MATH., UNIV. KOELN, REP. 75-3,
(1975)

431* MARZOUK, M.
INTERVALLANALYTISCHE METHODEN ZUR EINSCHLIESSUNG
DER EXTREMA VON FUNKTIONEN IN ZWEI VERAENDERLICHEN
DIPLOMARBEIT AM INST. F. ANGEW. MATH. U.
INFORMATIK, UNIVERSITAET BONN (1970)

432* MATULA, D.W.
IN-AND-OUT CONVERSIONS
COMMUNICATIONS OF THE ACM 11, 47-50 (1968)

MAYER, O.
SIEHE AUCH: ALEFELD, G.; HERZBERGER, J.;
MAYER, O.

433* MAYER, O.
DER SCHAUDERSCHE FIXPUNKTSATZ IN QUASILINEAREN
RAEUMEN
BER. D. RECHENZENTRUMS U. D. INST. F. ANGEW.
MATH., UNIVERSITAET KARLSRUHE (1968)

434* MAYER, O.
UEBER DIE IN DER INTERVALLRECHNUNG AUFTRETENDEN
RAEUME UND EINIGE ANWENDUNGEN
DISSERTATION, UNIVERSITAET KARLSRUHE (1968)

435* MAYER, O.
UEBER EIGENSCHAFTEN VON INTERVALLPROZESSEN
Z. ANGEW. MATH. MECH. 48, T86-T87 (1968)

436* MAYER, O.
UEBER EINE KLASSE KOMPLEXER
INTERVALLGLEICHUNGSSYSTEME MIT ITERATIONSFAEHIGER
GESTALT
COMPUTING 6, 104-106 (1970)

437* MAYER, O.
UEBER DIE BESTIMMUNG VON EINSCHLIESSUNGSMENGEN
FUER DIE LOESUNGEN LINEARER GLEICHUNGSSYSTEME MIT
FEHLERBEHAFTETEN KOEFFIZIENTEN
ANGEW. INFORMATIK (ELEKTRON. DATENVERARBEITUNG)
12, 164-167 (1970)

438* MAYER, O.
ALGEBRAISCHE UND METRISCHE STRUKTUREN IN DER
INTERVALLRECHNUNG UND EINIGE ANWENDUNGEN
COMPUTING 5, 144-162 (1970)

439* MAYER, O.
UEBER DIE INTERVALLMAESSIGE DURCHFUEHRUNG EINIGER
ITERATIONSVERFAHREN
Z. ANGEW. MATH. MECH. 50, T65-T66 (1970)

440* MAYER, O.
UEBER INTERVALLMAESSIGE ITERATIONSVERFAHREN BEI
LINEAREN GLEICHUNGSSYSTEMEN UND ALLGEMEINEREN
INTERVALLGLEICHUNGSSYSTEMEN
Z. ANGEW. MATH. MECH. 51, 117-124 (1971)

441* MEHLER, D.
EIN PROGRAMMIERSYSTEM MIT INTERVALLARITHMETIK FUER
DIE IBM 7090
BERICHT DES INST. F. INSTRUM. MATH. 47,
UNIVERSITAET BONN (1968)

442* MEINGUET, J.
LE CONTROLE DES ERREURS EN CALCUL AUTOMATIQUE
LABORATOIRE DE RECHERCHES, RAPPORT R 29,
MANUFACTURE BELGE DE LAMPES ET DE MATERIEL
ELECTRONIQUE, BRUXELLES (1965)

443* MEINGUET, J.
SENS ET PORTEE DES ARITHMETIQUES DE SIGNIFICATION
COLLOQUE INTERNATIONAL DU C.N.R.S., BESANCON
(1966)

444* MEINGUET, J.
ON THE ESTIMATION OF SIGNIFICANCE
'TOPICS IN INTERVAL ANALYSIS', ED. BY E. HANSEN,
OXFORD UNIVERSITY PRESS, 47-64 (1969)

445* MEISSNER, H.-P.
LOESUNG DES ERSTEN RANDWERTPROBLEMS EINER LINEAREN
GEWOEHNLICHEN DIFFERENTIALGLEICHUNG ZWEITER
ORDNUNG MIT FEHLERSCHRANKEN
DIPLOMARBEIT AM INST. F. INFORMATIK, UNIVERSITAET
KARLSRUHE (1969)

METROPOLIS, N.
SIEHE AUCH: ASHENHURST, R. L.; METROPOLIS, N.
                      GARDINER, V.; METROPOLIS, N.

446* METROPOLIS, N.; ASHENHURST, R.L.
SIGNIFICANT DIGIT COMPUTER ARITHMETIC
IRE TRANS. ELECTR. COMP. EC-7, 265-267 (1958)

447* MEYER, G.
ZUR NUMERISCHEN BEHANDLUNG VON LINEAREN
GLEICHUNGSSYSTEMEN MIT INTERVALLKOEFFIZIENTEN
DIPLOMARBEIT AM INST. F. ANGEW. MATH.,
UNIVERSITAET KARLSRUHE (1971)

448* MIEL, G.
ON A POSTERIORI ERROR ESTIMATES
MATH. COMPUT. 31, 204-213 (1977)

449* MIHELCIC, M.
EINE MODIFIKATION DES HALBIERUNGSVERFAHRENS ZUR
BESTIMMUNG ALLER REELLEN NULLSTELLEN EINER
FUNKTION MIT HILFE DER INTERVALL- ARITHMETIK
ANGEW. INFORMATIK (ELEKTRON. DATENVERARBEITUNG)
17, 25-29 (1975)

450* MILLER, R. H.
AN EXAMPLE IN 'SIGNIFICANT-DIGIT' ARITHMETIC
COMM. ACM 7, 21 (1964)

MILLER, W.
SIEHE AUCH: CHUBA, W.; MILLER, W.

MILLER-MOORE

451* MILLER, W.
ON AN INTERVAL-ARITHMETIC MATRIX METHOD
BIT 12, 213-219 (1972)

452* MILLER, W.
QUADRATIC CONVERGENCE IN INTERVAL ARITHMETIC,
PART II
BIT 12, 291-298 (1972)

453* MILLER, W.
AUTOMATIC A PRIORI ROUND-OFF ANALYSIS
(PART I: ARITHMETIC CONDITIONS)
COMPUTING 10, 97-106 (1972)

454* MILLER, W.
MORE ON QUADRATIC CONVERGENCE IN INTERVAL
ARITHMETIC
BIT 13, 76-83 (1973)

455* MILLER, W.
THE ERROR IN INTERVAL ARITHMETIC
IBM THOMAS J. WATSON RESEARCH CENTER, RC 4338
(#19410), YORKTOWN HEIGHTS, NEW YORK (1973)

456* MILLER, W.
THE ERROR IN INTERVAL ARITHMETIC
'INTERVAL MATHEMATICS', ED. BY K. NICKEL
LECTURE NOTES IN COMPUTER SCIENCE 29, SPRINGER
VERLAG, 246-250 (1975)

457* MILLER, W.
AUTOMATIC VERIFICATION OF NUMERICAL STABILITY
ERSCHEINT IN COMM. ACM

458* MILLER, W.
GRAPH TRANSFORMATIONS FOR ROUNDOFF ANALYSIS
SIAM J. COMPUT. 5, 204-216 (1976)

459* MILLER, W.; PAULHAMUS, R.
AUTOMATIC A PRIORI ROUND-OFF ANALYSIS

(PART II: SYMBOL MANIPULATION CONDITIONS)
TECHNICAL REPORT, PENNSYLVANIA STATE UNIVERSITY
(1972)

460* MIRKIN, B. G.
DESCRIPTIONS OF SOME RELATIONS ON THE SET OF
REAL-LINE INTERVALS
J. MATH. PSYCHOLOGY 9, 243-252 (1972)

MITTAL, B.S.
SIEHE AUCH: LATA, M.; MITTAL, B.S.

461* MOENCH, W.
MONOTONE EINSCHLIESSUNG VON POSITIVEN INVERSEN
Z. ANGEW. MATH. MECH. 53, 207-208 (1973)

462* MOENCH, W.
MONOTONE EINSCHLIESSUNG VON LOESUNGEN
NICHTLINEARER GLEICHUNGEN
DISSERTATION, TECHN. UNIVERSITAET DRESDEN (1973)

463* MOENCH, W.
MONOTONE EINSCHLIESSUNG VON LOESUNGEN
NICHTLINEARER GLEICHUNGSSYSTEME DURCH UEBERLINEAR
KONVERGENTE MEHRSCHRITTVERFAHREN
Z. ANGEW. MATH. MECH. 54, 807-812 (1974)

MONTGOMERY, D.
SIEHE: ELLENBERG, S.; MONTGOMERY, D.

MOORE, R.E.
SIEHE AUCH: BRAUN, J.A.; MOORE, R.E.
            DANIEL, J.W.; MOORE, R.E.
            GORN, S.; MOORE, R.E.
            LEE, Y.D.; MOORE, R.E.

464* MOORE, R. E.
AUTOMATIC ERROR ANALYSIS IN DIGITAL COMPUTATION
TECHNICAL REPORT LMSD-48421, LOCKHEED MISSILES AND
SPACE DIVISION, SUNNYVALE, CAL. (1959)

465* MOORE, R. E.
INTERVAL ARITHMETIC AND AUTOMATIC ERROR ANALYSIS
IN DIGITAL COMPUTING
PH. D. THESIS, APPLIED MATHEMATICS AND STATISTICS
LABORATORIES, REPORT 25, STANFORD UNIVERSITY
(1962)

466* MOORE, R.E.
THE AUTOMATIC ANALYSIS AND CONTROL OF ERROR IN
DIGITAL COMPUTATION BASED ON THE USE OF INTERVAL
NUMBERS
´ERROR IN DIGITAL COMPUTATION, VOL. I´ ED. BY
L.B.RALL, WILEY & SONS INC, NEW YORK, 61-130
(1965)

467* MOORE, R. E.
AUTOMATIC LOCAL COORDINATE TRANSFORMATIONS TO
REDUCE THE GROWTH OF ERROR BOUNDS IN INTERVAL
COMPUTATION OF SOLUTIONS OF ORDINARY DIFFERENTIAL
EQUATIONS
´ERROR IN DIGITAL COMPUTATION, VOL II´ ED. BY
L.B.RALL, WILEY & SONS INC, NEW YORK, 103-140
(1965)

468* MOORE, R.E.
INTERVAL ANALYSIS
PRENTICE-HALL, INC., ENGLEWOOD CLIFFS, N.J.
(1966)

469* MOORE, R.E.
PRACTICAL ASPECTS OF INTERVAL COMPUTATION
APL. MAT. 13, 52-92 (1968)

470* MOORE, R. E.
FUNCTIONAL ANALYSIS FOR COMPUTERS
´FUNKTIONALANALYTISCHE METHODEN DER NUMERISCHEN
MATHEMATIK´, ISNM 12, HERAUSG. L. COLLATZ, H.
UNGER, BIRKHAEUSER VERLAG, STUTTGART, BASEL,
113-126 (1969)

471* MOORE, R.E.
INTERVALLANALYSE
DEUTSCHE UEBERSETZUNG VON: MOORE, R. E., INTERVAL
ANALYSIS R. OLDENBOURG VERLAG, MUENCHEN, WIEN
(1969)

472* MOORE, R. E.
INTRODUCTION TO ALGEBRAIC PROBLEMS
´TOPICS IN INTERVAL ANALYSIS´, ED. BY E. HANSEN,
OXFORD UNIVERSITY PRESS, 3-9 (1969)

473* MOORE, R.E.
INTRODUCTION TO CONTINUOUS PROBLEMS
´TOPICS IN INTERVAL ANALYSIS´, ED. BY E. HANSEN,
OXFORD UNIVERSITY PRESS, 67-73 (1969)

474* MOORE, R. E.
TWO-SIDED APPROXIMATION TO SOLUTIONS OF NONLINEAR
OPERATOR EQUATIONS - A COMPARISON OF METHODS
FROM CLASSICAL ANALYSIS, FUNCTIONAL ANALYSIS,
AND INTERVAL ANALYSIS
´INTERVAL MATHEMATICS´, ED. BY K. NICKEL
LECTURE NOTES IN COMPUTER SCIENCE 29, SPRINGER
VERLAG, 31-47 (1975)

475* MOORE, R.E.
MATHEMATICAL ELEMENTS OF SCIENTIFIC COMPUTING
TO APPEAR: HOLT, RINEHART AND WINSTON, INC. NEW
YORK

476* MOORE, R. E.
ON COMPUTING THE RANGE OF A RATIONAL FUNCTION OF N
VARIABLES OVER A BOUNDED REGION
INTERNER BERICHT DES INST.F. PRAKT. MATH. 75/2,
UNIVERSITAET KARLSRUHE (1975)
PUBLIZIIERT IN:
COMPUTING 16, 1-15 (1976)

477* MOORE, R.E.
BOUNDING SETS
SUBMITTED TO SIAM REV. 2, (1976)

478* MOORE, R.E.
A TEST FOR EXISTENCE OF SOLUTIONS TO NONLINEAR
SYSTEMS
SIAM J. NUMER. ANAL., (TO APPEAR)

MOORE-NICKEL

479* MOORE, R.E.; DAVIDSON, J.A.; JASCHKE, H.R.;
SHAYER, S.
DIFEQ INTEGRATION ROUTINE-USER'S MANUAL
TECHNICAL REPORT: MATHEMATICS, LMSC 6-90-64-6,
LOCKHEED MISSILES AND SPACE CO., PALO ALTO, CAL.
(1964)

480* MOORE, R.E.; STROTHER, W.; YANG, C.T.
INTERVAL INTEGRALS
TECHNICAL MEMORANDUM: MATHEMATICS, LMSD-703073,
LOCKHEED MISSILES AND SPACE DIVISION, SUNNYVALE,
CAL. (1960)

481* MOORE, R. E.; YANG, C. T.
INTERVAL ANALYSIS I
TECHNICAL DOCUMENT, LMSD-285875, LOCKHEED MISSILES
AND SPACE DIVISION, SUNNYVALE, CAL. (1959)

482* MOR, M.
ON THE THREE-VALUED SIMULATION OF DIGITAL SYSTEMS
IEEE TRANS. COMPUTERS C 25, 1152-1156 (1976)

483* MORLOCK, M.
UEBER DAS NEWTON-SIMULTANVERFAHREN UND SEINE
ANWENDUNG AUF DIE ABGEBROCHENE EXPONENTIALREIHE
DIPLOMARBEIT AM INST. F. INFORMATIK, UNIVERSITAET
KARLSRUHE (1969)

MUENZER, W.
SIEHE: CSAJKA, I.; MUENZER, W.; NICKEL, K.

484* MUHAMMAD, G.
INTERVAL ARITHMETIC
DISSERTATION FOR M. SC. MATHEMATICS, HERTFORD
COLLEGE, OXFORD (1975)

485* NADLER, S. B. JR.
MULTI-VALUED CONTRACTION MAPPINGS (ABSTRACT)
NOTICES OF AMER MATH. SOC. 14, 930 (1967)

486* NADLER, S. B. JR.
MULTI-VALUED CONTRACTION MAPPINGS
PACIFIC J. MATH. 30, 475-488 (1969)

487* NAGEL, K.-D.
EIN VERSUCH ZUR BESTIMMUNG VON OPTIMALEN
ZWISCHENSTELLEN FUER QUADRATURVERFAHREN MIT
FEHLERSCHRANKEN
DIPLOMARBEIT AM INST. F. PRAKT. MATH. DER
UNIVERSITAET KARLSRUHE (1977)

NAUBER, W.
SIEHE AUCH: SCHMIDT, J.W.; NAUBER, W.

488* NEHRKORN,K.
METHODEN DER REALISIERUNG EINER REELLEN
INTERVALLARITHMETIK IN EINEM FORTRAN-SYSTEM
WISS. Z. TECH. HOCHSCH. KARL-MARX-STADT 16,
469-474 (1974)

489* NEULAND, W.
DIE NUMERISCHE BEHANDLUNG VON INTEGRALGLEICHUNGEN
MIT INTERVALL-ANALYTISCHEN METHODEN
BERICHTE DER GMD BONN 21 (1969)

490* NEULAND, W.
DIE INTERVALLARITHMETISCHE BEHANDLUNG SPEZIELLER
FREDHOLMSCHER INTEGRALGLEICHUNGEN
Z. ANGEW. MATH. MECH. 50, T69 (1970)

NICKEL, K.
SIEHE AUCH: APOSTOLATOS, N.; KULISCH, U.;
NICKEL, K.
CSAJKA,I.; MUENZER, W.; NICKEL, K.
ROBINSON, S.M.; NICKEL, K.
WIPPERMANN, H.-W. ET AL.

491* NICKEL, K.
UEBER DIE NOTWENDIGKEIT EINER
FEHLERSCHRANKENARITHMETIK FUER RECHENAUTOMATEN
NUMER. MATH. 9, 69-79 (1966)

535* OLIVEIRA, F.A.
ALGUNS ASPECTOS DO PROBLEMA DOS ERROS EM ANALISE
NUMERICA
GAZ. MAT. 29, 11-14, 109-112 (1968)

536* OLIVEIRA, F.A.
INTERVAL ANALYSIS AND TWO-POINT-BOUNDARY-VALUE
PROBLEMS
SUMMARY REPORT FAO-70-L, UNIVERSITY OF COIMBRA,
PORTUGAL (1970)

537* OLIVEIRA, F.A.A.N. DE
SOBRE A APLICACAO DA ANALISE INTERVALAR A
RESOLUCAO NUMERICA DE EQUACOES DIFERENCIAIS
DISSERTACAO PARA DOUTORAMENTO EM MATEMATICA PURA
NA FACULDADE DE CIENCIAS DA UNIVERSIDADE DE
COIMBRA (1970)

538* OLIVEIRA, F.A.
ERROR ANALYSIS AND COMPUTER ARITHMETIC
SUMMARY REPORT FAO-2, UNIVERSITY OF COIMBRA,
PORTUGAL (1971)

539* OLIVEIRA, F. A.
INTERVAL ANALYSIS AND TWO-POINT BOUNDARY VALUE
PROBLEMS
SIAM J. NUMER. ANAL. 11, 382-391 (1974)

540* OLIVEIRA, F. A.
INTERVAL-EXTENSION OF QUASILINEARIZATION METHOD
'INTERVAL MATHEMATICS', ED. BY K. NICKEL
LECTURE NOTES IN COMPUTER SCIENCE 29, SPRINGER
VERLAG, 270-278 (1975)

541* OLVER, F.W.J.
A NEW APPROACH TO ERROR ARITHMETIC
ERSCHEINT IN SIAM JOURNAL ON NUMERICAL ANALYSIS

542* ORTOLF, H.-J.
EINE VERALLGEMEINERUNG DER INTERVALLARITHMETIK
BERICHTE DER GMD BONN 11, (1969)

543* OST, F.
SIEHE AUCH: FISCHER, H.; OST, F.

544* OST, F.
OPERATIONEN UND LINEARE ABBILDUNGEN IN
HYPERNORMIERTEN RAEUMEN
DIPLOMARBEIT AM INST. F. STATISTIK UND
UNTERNEHMENSFORSCHUNG, TECHN. UNIVERSITAET
MUENCHEN (1975)

545* OSWALD, W.
BESCHREIBUNG DER PROGRAMMIERSPRACHE TRIPLEX-ALGOL
UND ERSTELLUNG EINES COMPILERS IN ALGOL 60
DIPLOMARBEIT AM INST. F. INFORMATIK, UNIVERSITAET
KARLSRUHE (1969)

546* PARKER, D. S., JR.
THE STATISTICAL THEORY OF RELATIVE ERRORS IN
FLOATING-POINT COMPUTATION
REPORT NO. UIUCDCS-R-76-787 DEPARTEMENT OF
COMPUTER SCIENCE UNIVERSITY OF ILLINOIS AT
URBANA-CHAMPAIGN, URBANA, ILLINOIS, (1976)

547* PAULHAMUS, R.
SIEHE: MILLER, W.; PAULHAMUS, R.

548* PLATZOEDER, L.
UEBER OPTIMALE ITERATIONSVERFAHREN BEI KOMPLEXEN
GLEICHUNGSSYSTEMEN
DIPLOMARBEIT AM INST. F. ANGEW. MATH.,
UNIVERSITAET KARLSRUHE (1974)

549* PLATZOEDER, L.
UEBER DIE MIT HILFE DER INTERVALLARITHMETIK
DURCHGEFUEHRTEN ITERATIONSVERFAHREN ZUR
LOESUNGSEINSCHLIESSUNG BEI LINEAREN
GLEICHUNGSSYSTEMEN
Z. ANGEW. MATH. MECH. 56, T308-T309 (1976)

PRAGER, W.
SIEHE: OETTLI, W.; PRAGER, W.
OETTLI, W.; PRAGER, W.; WILKINSON, J.H.

520* NICKEL, K.
INTERVAL-ANALYSIS
THE STATE OF THE ART IN NUMERICAL ANALYSIS,
JACOBS, D., PROCEEDINGS OF THE CONFERENCE ON THE
STATE OF THE ART IN NUMERICAL ANALYSIS HELD AT THE
UNIVERSITY OF YORK, APRIL 12TH-15TH, 1976, 193-226
(1977)

521* NICKEL, K.
AUFGABEN VON MONOTONER ART UND
INTERVALL-MATHEMATIK
ERSCHEINT IN Z. ANGEW. MATH. MECH.

522* NICKEL, K.
EIN ZUSAMMENHANG ZWISCHEN AUFGABEN MONOTONER ART
UND INTERVALL-MATHEMATIK
ERSCHEINT DEMNAECHST

523* NICKEL, K.; RITTER, K.
TERMINATION CRITERION AND NUMERICAL CONVERGENCE
SIAM J. NUMER. ANAL. 9, 277-283 (1972)

NUDING, E.
SIEHE AUCH: BARTH, W.; NUDING, E.

524* NUDING, E.
INNERE LOESUNGEN VON INTERVALLGLEICHUNGEN
TAGUNG UEBER INTERVALLRECHNUNG IN OBERWOLFACH,
VORTRAGSMANUSKRIPT (1972)

525* NUDING, E.
INTERVALLARITHMETIK, TEIL 1
SKRIPTUM ZUR VORLESUNG, RECHENZENTRUM DER
UNIVERSITAET HEIDELBERG (1972)

526* NUDING, E.
INTERVALLARITHMETIK, TEIL 2
SKRIPTUM ZUR VORLESUNG, RECHENZENTRUM DER
UNIVERSITAET HEIDELBERG (1972)

527* NUDING, E.
OPTIMALLOESUNGEN VON INTERVALL-GLEICHUNGSSYSTEMEN

INTERNER BERICHT D. INST. F. PRAKT. MATH. 73/6,
UNIVERSITAET KARLSRUHE (1973)

528* NUDING, E.
INTERVALLRECHNUNG UND WIRKLICHKEIT
"INTERVAL MATHEMATICS", ED. BY K. NICKEL
LECTURE NOTES IN COMPUTER SCIENCE 29, SPRINGER
VERLAG, 263-269 (1975)

529* NUDING, E.; WARMBOLD, I.
VIERPOLTHEORIE UND TRANSMISSIONSMATRIX-VERFAHREN
IN DER REAKTORBERECHNUNG
Z. ANGEW. MATH. MECH. 47, T69-T72 (1967)

530* NUDING, E; WILHELM, J.
UEBER GLEICHUNGEN UND UEBER LOESUNGEN
Z. ANGEW. MATH. MECH. 52, T188-T190 (1972)

OELSCHLAEGEL, D.
SIEHE AUCH: SUESSE, H.; OELSCHLAEGEL, D.

531* OETTLI, W.
ON THE SOLUTION SET OF A LINEAR SYSTEM WITH
INACCURATE COEFFICIENTS
SIAM J. NUMER. ANAL. 2, 115-118 (1965)

532* OETTLI, W.; PRAGER, W.
COMPATIBILITY OF APPROXIMATE SOLUTION OF LINEAR
EQUATIONS WITH GIVEN ERROR BOUNDS FOR COEFFICIENTS
AND RIGHT-HAND SIDES
NUMER. MATH. 6, 405-409 (1964)

533* OETTLI, W.; PRAGER, W.; WILKINSON, J.H.
ADMISSIBLE SOLUTIONS OF LINEAR SYSTEMS WITH NOT
SHARPLY DEFINED COEFFICIENTS
SIAM J. NUMER. ANAL. 2, 291-299 (1965)

534* OHNAKA, K.; YASUI, H.; KUBO, T.
AN AUTOMATIC ERROR ESTIMATION METHOD IN NUMERICAL
CALCULATION
TECHNOL. REP. OSAKA UNIV. (JAPAN), 25, 257-265
(1975)

507* NICKEL, K.
FEHLERSCHRANKEN FUER DIE NUMERISCH BERECHNETEN
WURZELN EINES POLYNOMS (NOTIZ)
Z. ANGEW. MATH. MECH. 50, T81 (1970)

508* NICKEL, K.
THE PRAE-EULER SUMMATION METHOD
MRC TECHNICAL SUMMARY REPORT # 1072 UNIVERSITY OF
WISCONSIN, MADISON (1971)
UND (DEUTSCHE UEBERSETZUNG):
DAS PRAE-EULER'SCHE LIMITIERUNGSVERFAHREN
INTERNER BER. DES INST. F. INFORMATIK 70/7,
UNIVERSITAET KARLSRUHE (1970)

509* NICKEL, K.
ON THE NEWTON METHOD IN INTERVAL ANALYSIS
MRC TECHNICAL SUMMARY REPORT # 1136 UNIVERSITY OF
WISCONSIN, MADISON (1971)

510* NICKEL, K.
THE CONTRACTION MAPPING FIXED POINT THEOREM IN
INTERVAL ANALYSIS
MRC TECHNICAL SUMMARY REPORT # 1334 UNIVERSITY OF
WISCONSIN, MADISON (1973)

511* NICKEL, K.
STABILITY AND CONVERGENCE OF MONOTONIC ALGORITHMS
MRC TECHNICAL SUMMARY REPORT # 1340 UNIVERSITY OF
WISCONSIN, MADISON (1973)

512* NICKEL, K.
UEBER DIE STABILITAET UND KONVERGENZ
NUMERISCHER ALGORITHMEN, TEIL I INTERNER BERICHT
DES INST. F. PRAKT. MATH. 74/3, UNIVERSITAET
KARLSRUHE (1974)
PUBLIZIERT IN:
COMPUTING 15, 291-309 (1975)

513* NICKEL, K.
UEBER DIE STABILITAET UND KONVERGENZ
NUMERISCHER ALGORITHMEN, TEIL II INTERNER BERICHT
DES INST. F. PRAKT. MATH. 74/4, UNIVERSITAET
KARLSRUHE (1974)
PUBLIZIERT IN:
COMPUTING 15, 311-328 (1975)

514* NICKEL, K.
INTERVAL MATHEMATICS
LECTURE NOTES IN COMPUTER SCIENCE 29, SPRINGER
VERLAG (1975)

515* NICKEL, K.
INTERVAL-MATHEMATIK. ZUM KARLSRUHER SYMPOSIUM
1975.
`INTERVAL MATHEMATICS`, ED. BY K. NICKEL
LECTURE NOTES IN COMPUTER SCIENCE 29, SPRINGER
VERLAG, 1-6 (1975)

516* NICKEL, K.
VERBANDSTHEORETISCHE GRUNDLAGEN DER
INTERVALL-MATHEMATIK
`INTERVAL MATHEMATICS`, ED. BY K. NICKEL
LECTURE NOTES IN COMPUTER SCIENCE 29, SPRINGER
VERLAG, 251-262 (1975)

517* NICKEL, K.
DIE UEBERSCHAETZUNG DES WERTEBEREICHS EINER
FUNKTION IN DER INTERVALLRECHNUNG MIT ANWENDUNGEN
AUF LINEARE GLEICHUNGSSYSTEME
INTERNER BERICHT DES INST. F. PRAKT. MATH. 75/5,
UNIVERSITAET KARLSRUHE (1975)
PUBLIZIERT IN:
COMPUTING 18, 15-36 (1977)

518* NICKEL, K.L.
STABILITY AND CONVERGENCE OF MONOTONIC ALGORITHMS
JOURN. OF MATH. ANAL. AND APPL. 54, 157-172 (1976)

519* NICKEL, K.
INTERVAL-ANALYSIS
PROCEEDINGS DER TAGUNG DES IMA UEBER `THE STATE
OF-THE-ART IN NUMERICAL ANALYSIS' YORK/ENGLAND
(1976)

492* NICKEL, K.:
DIE NUMERISCHE BERECHNUNG DER WURZELN EINES
POLYNOMS
NUMER. MATH. 9, 80-98 (1966)

493* NICKEL, K.:
DIE BERECHNUNG VON POLYNOMWURZELN (NOTIZ)
Z. ANGEW. MATH. MECH. 46, T74 (1966)

494* NICKEL, K.:
ZWEI NEUE RECHENMASCHINEN-SYSTEME AN DER TH
KARLSRUHE: HYDRA-X8 UND TRIPLEX-ALGOL-Z23
UMSCHAU 67, 525-526 (1967)

495* NICKEL, K.:
DIE VOLLAUTOMATISCHE BERECHNUNG EINER EINFACHEN
NULLSTELLE VON F(T)=0 EINSCHLIESSLICH EINER
FEHLERABSCHAETZUNG
COMPUTING 2, 232-245 (1967)

496* NICKEL, K.:
QUADRATURVERFAHREN MIT FEHLERSCHRANKEN
Z. ANGEW. MATH. MECH. 47, T68-T69 (1967)

497* NICKEL, K.:
ANWENDUNGEN EINER FEHLERSCHRANKEN-ARITHMETIK
'NUMERISCHE MATHEMATIK, DIFFERENTIALGLEICHUNGEN,
APPROXIMATIONSTHEORIE', ISNM 9, HERAUSG.
L.COLLATZ, G. MEINARDUS, H. UNGER, BIRKHAEUSER
VERLAG STUTTGART, BASEL, 285-304 (1968)

498* NICKEL, K.:
QUADRATURVERFAHREN MIT FEHLERSCHRANKEN
COMPUTING 3, 47-64 (1968)

499* NICKEL, K.:
EIN STETS KONVERGENTER NEWTON-ALGORITHMUS MIT
FEHLERABSCHAETZUNG (NOTIZ)
Z. ANGEW. MATH. MECH. 48, T111 (1968)

500* NICKEL, K.:
BERICHT UEBER NEUE KARLSRUHER ERGEBNISSE BEI DER
FEHLERFASSUNG VON NUMERISCHEN PROZESSEN
APL. MAT. 13, 168-173 (1968)

501* NICKEL, K.:
ERROR BOUNDS AND COMPUTER-ARITHMETIC
'PROC. OF IFIP-CONGRESS 1968, VOL. I', ED. BY
A.J.H.MORRELL, NORTH-HOLLAND PUBL. COMP. AMSTERDAM,
54-60 (1969) VGL.: KAHAN, W. M., INVITED
COMMENTARY, EBENDA, 60-62

502* NICKEL, K.:
THE APPLICATION OF INTERVAL ANALYSIS TO THE
NUMERICAL SOLUTION OF DIFFERENTIAL EQUATIONS
INTERNER BERICHT DES INST. F. INFORMATIK 69/9,
UNIVERSITAET KARLSRUHE (1969)

503* NICKEL, K.:
TRIPLEX-ALGOL AND APPLICATIONS
'TOPICS IN INTERVAL ANALYSIS', ED. BY E. HANSEN,
OXFORD UNIVERSITY PRESS, 10-24 (1969)

504* NICKEL, K.:
ZEROS OF POLYNOMIALS AND OTHER TOPICS
'TOPICS IN INTERVAL ANALYSIS', ED. BY E. HANSEN,
OXFORD UNIVERSITY PRESS, 25-34 (1969)

505* NICKEL, K.:
DAS KAHAN-BABUSKA'SCHE SUMMIERUNGSVERFAHREN IN
TRIPLEX-ALGOL 60
INTERNER BERICHT DES INST. F. INFORMATIK 69/3,
UNIVERSITAET KARLSRUHE (1969)
PUBLIZIERT IN:
Z. ANGEW. MATH. MECH. 50, 369-373 (1970)

506* NICKEL, K.:
FEHLERSCHRANKEN ZU NAEHERUNGSWERTEN VON
POLYNOMWURZELN
INTERNER BERICHT DES INST. F. INFORMATIK 69/6,
UNIVERSITAET KARLSRUHE (1969)
PUBLIZIERT IN:
COMPUTING 6, 9-27 (1970)

548* PRESTELE, H.
SPLINEFUNKTIONEN IN DER INTERVALLARITHMETIK
DIPLOMARBEIT AM INST. F. ANGEW. MATH., TECHNISCHE
UNIVERSITAET MUENCHEN (1972)

RABINOWITZ, P.
SIEHE AUCH: DAVIS, P.; RABINOWITZ, P.

549* RADSTROEM, H.
AN EMBEDDING THEOREM FOR SPACES OF CONVEX SETS
PROC. AMER. MATH. SOC. 3, 165-169 (1952)

550* RAITH, M.
DIE NUMERISCHE BEHANDLUNG EINES RANDWERTPROBLEMS
FUER EINE GEWOEHNLICHE DIFFERENTIALGLEICHUNG
ZWEITER ORDNUNG MIT FEHLERERFASSUNG
DIPLOMARBEIT AM INST. F. PRAKT. MATH.,
UNIVERSITAET KARLSRUHE (1976)

RALL, L.B.
SIEHE: GRAY, J.H.; RALL, L.B.
        KUBA, D.; RALL, L.B.

551* RATNER, L.
MULTI-VALUED TRANSFORMATIONS
UNIVERSITY OF CALIFORNIA (1949)

RATSCHEK, H.
SIEHE AUCH: ECKER, K.; RATSCHEK, H.

552* RATSCHEK, H.
UBER EINIGE INTERVALLARITHMETISCHE GRUNDBEGRIFFE
COMPUTING 4, 43-55 (1969)

553* RATSCHEK, H.
DIE BINAEREN SYSTEME DER INTERVALLARITHMETIK
COMPUTING 6, 295-308 (1970)

554* RATSCHEK, H.
DIE SUBDISTRIBUTIVITAET DER INTERVALLARITHMETIK
Z. ANGEW. MATH. MECH. 51, 189-192 (1971)

555* RATSCHEK, H.
TEILBARKEITSGESETZE DER INTERVALLARITHMETIK
Z. ANGEW. MATH. MECH. 51, 170-171 (1971)

556* RATSCHEK, H.
TEILBARKEITSKRITERIEN DER INTERVALLARITHMETIK
J. REINE ANGEW. MATH. 252, 128-138 (1972)

557* RATSCHEK, H.
ERGEBNISSE EINER UNTERSUCHUNG UEBER DIE STRUKTUR
VON INTERVALLPOLYNOMEN
BERICHTE DER GMD BONN 52 (1972)

558* RATSCHEK, H.
GLEICHHEIT VON PRODUKT UND FORMALPRODUKT BEI
INTERVALLPOLYNOMEN
COMPUTING 10, 245-254 (1972)

559* RATSCHEK, H.
INTERVALLARITHMETIK - MIT ZIRKEL UND LINEAL
ELEM. MATH. 28, 93-96 (1973)

560* RATSCHEK, H.
UEBER EINIGE WESENSZUEGE DER INTERVALLMATHEMATIK
MATH.-PHYS. SEMESTERBER. XXI, 67-79 (1974)

561* RATSCHEK, H.
MITTELWERTSATZ FUER INTERVALLFUNKTIONEN
Z. ANGEW. MATH. MECH. 54, T229-T230 (1974)

562* RATSCHEK, H.
UEBER DAS PRODUKT VON INTERVALLPOLYNOMEN
COMPUTING 13, 313-325 (1974)

563* RATSCHEK, H.
NICHTNUMERISCHE ASPEKTE DER INTERVALLARITHMETIK
"INTERVAL MATHEMATICS", ED. BY K. NICKEL
LECTURE NOTES IN COMPUTER SCIENCE 29, SPRINGER
VERLAG, 48-74 (1975)

564* RATSCHEK, H.
MITTELWERTSAETZE FUER INTERVALLFUNKTIONEN
MANUSKRIPT

565* RATSCHEK, H.
CENTRED FORMS
VORTRAG IN OBERWOLFACH (1976)
INTERVALL-MATHEMATIK-TAGUNG

566* RATSCHEK, H.
FEHLERERFASSUNG MIT PARTIELLEN MENGEN
COMPUTING SUPPLEMENTUM 1, ALBRECHT, R.;
KULISCH, U., SPRINGER-VERLAG WIEN, NEW YORK,
121-128 (1977)

567* RATSCHEK, H.; SCHROEDER, G.
UEBER DIE ABLEITUNG VON INTERVALLWERTIGEN
FUNKTIONEN
COMPUTING 7, 172-187 (1971)

568* RATSCHEK, H.; SCHROEDER, G.
UEBER DIE DIFFERENTIATION VON INTERVALLFUNKTIONEN
Z. ANGEW. MATH. MECH. 52, T219-T220 (1972)

569* RATSCHEK, H.; SCHROEDER, G.
UEBER DEN QUASILINEAREN RAUM
BERICHT NR. 65 (1976) MATH. STAT. SEKTION,
FORSCHUNGSZENTRUM GRAZ

570* RATSCHEK, H.; SCHROEDER, G.
DARSTELLUNG VON HALBGRUPPEN ALS SYSTEME
ABGESCHLOSSENER BESCHRAENKTER KONVEXER MENGEN
ERSCHEINT DEMNAECHST

571* RAUER, H.J.
EINE TRIPLEX-PROZEDUR ZUR AUSWERTUNG DES
VOLLSTAENDIGEN ELLIPTISCHEN INTEGRALS DRITTER
GATTUNG
INTERNER BERICHT DES INST. F. INFORMATIK 69/1,
UNIVERSITAET KARLSRUHE (1969)

572* REDFIELD, R.H.
GENERALIZED INTERVALS AND TOPOLOGY
CZECHOSLOVAK MATH. J. 26, 527-540 (1976)

REEVES, T. E.
SIEHE: BREITER, M.C.; KELLER, C. L.;
REEVES, T. E.;
KELLER, C. L.; REEVES, T. E.

573* REICHMANN, K.
DIE NUMERISCHE BEHANDLUNG VON LINEAREN
GLEICHUNGSSYSTEMEN MIT TRIDIAGONALMATRIZEN MIT
RUNDUNGSFEHLERERFASSUNG
STAATSEXAMENSARBEIT AM INST. F. PRAKT. MATH.,
UNIVERSITAET KARLSRUHE (1976)

574* REIF, N.
EINSCHLIESSUNGSVERFAHREN ZUR BESTIMMUNG DER
LOESUNGEN EINES NICHTLINEAREN GLEICHUNGSSYSTEMS
DIPLOMARBEIT AM INST. F. ANGEW. MATH. U.
INFORMATIK, UNIVERSITAET BONN (1973)

REINHART, E.
SIEHE AUCH: HEINDL, G.; REINHART, E.

575* REINHART, E.
EXAKTE ABSCHAETZUNG VON MAXIMALFEHLERN AUS
VORGEGEBENEN TOLERANZEN DER BEOBACHTUNGSGROESSEN
DEUTSCHE GEODAET. KOMMISSION BEI D. BAYER. AKAD.
D. WISS. REIHE C: DISSERTATIONEN HEFT NR. 211
(1975)

576* REITER, A.
INTERVAL ARITHMETIC PACKAGE (INTERVAL) FOR THE CDC
1604 AND CDC 3600
MRC TECHNICAL SUMMARY REPORT # 794 UNIVERSITY OF
WISCONSIN, MADISON (1968)

577* REITER, A.
INTERVAL ARITHMETIC PACKAGE; MRC PROGRAM 2
COOP ORGAN, CODE-WISC. MATH. RES. CENTER,
UNIVERSITY OF WISCONSIN, MADISON

578* RICHMAN, P.L.
VARIABLE-PRECISION INTERVAL ARITHMETIC TECHNICAL
MEMORANDUM MM 69-1374-26
BELL TELEPHONE LABORATORIES, INC., MURRAY HILL,
NEW JERSEY (1969)

579* RICHMAN, P.L.
ERROR CONTROL AND THE MIDPOINT PHENOMENON
TECHNICAL MEMORANDUM MM 69-1374-29
BELL TELEPHONE LABORATORIES, INC., MURRAY HILL,
NEW JERSEY (1969)

580* RICHMAN, P. L.
AUTOMATIC ERROR ANALYSIS FOR DETERMINING PRECISION
COMM. ACM 15, 813-817 (1972)

581* RICHMAN, P. L.
VARIABLE-PRECISION EXPONENTIATION
COMM. ACM 16, 38-40 (1973)

582* RICHMAN, P.L.
COMPUTING A SUBINTERVAL OF THE IMAGE
J. ASSOC. COMPUT. MACH. 21, 454-458 (1974)

583* RICHMAN, P.L.; GOLDSTEIN, A.J.
A MIDPOINT PHENOMENON
J. ASSOC. COMPUT. MACH. 20, 301-304 (1973)

584* RICHTMEYER, R. D.
THE ESTIMATION OF SIGNIFICANCE
AEC RESEARCH AND DEVELOPMENT REPORT, NYO-9083
(1960)

585* RIEDER, P.
EIN- UND AUSGABE
INTERNER BERICHT DES INST. F. INFORMATIK 70/5,
UNIVERSITAET KARLSRUHE (1970)

586* RIGAL, J. L.; GACHES, J.
ON THE COMPATIBILITY OF A GIVEN SOLUTION WITH THE
DATA OF A LINEAR SYSTEM
J. ASSOC. COMPUT. MACH. 14, 543-548 (1967)

587* RIS, F. N.
INTERVAL ANALYSIS AND APPLICATIONS TO LINEAR
ALGEBRA
D. PHIL. THESIS, OXFORD (1972)

588* RIS, F. N.
ON THE COMPUTATION OF GUARANTEED UPPER BOUNDS TO
THE SPECTRAL RADII OF NON-NEGATIVE MATRICES
IBM THOMAS J. WATSON RESEARCH CENTER YORKTOWN
HEIGHTS, NEW YORK (1974)

589* RIS, F. N.
COMPUTING RIGOROUS UPPER BOUNDS TO THE SPECTRAL
RADII OF NON-NEGATIVE MATRICES
INFORMATION PROCESSING 74 - NORTH HOLLAND
PUBLISHING COMPANY (1974)

590* RIS, F. N.
TOOLS FOR THE ANALYSIS OF INTERVAL ARITHMETIC
'INTERVAL MATHEMATICS', ED. BY K. NICKEL
LECTURE NOTES IN COMPUTER SCIENCE 29, SPRINGER
VERLAG, 75-98 (1975)

591* RIS, F.N.
A UNIFIED DECIMAL FLOATING-POINT ARCHITECTURE FOR
THE SUPPORT OF HIGH-LEVEL LANGUAGES (EXTENDED
ABSTRACT)
RESEARCH REPORT RC 6203 (#26651) 9/14/76 IBM
THOMAS J. WATSON RESEARCH CENTER YORKTOWN HEIGHTS,
NEW YORK 10598

RITTER, K.
SIEHE AUCH: NICKEL, K.; RITTER, K.

ROBERS, P.D.
SIEHE AUCH: BEN-ISRAEL, A.; ROBERS, P.D.

592* ROBERS, P.D.; BEN-ISRAEL, A.
A SUBOPTIMIZATION METHOD FOR INTERVAL LINEAR
PROGRAMMING: A NEW METHOD FOR LINEAR PROGRAMING
LINEAR ALGEBRA APPL. 3, 383-405 (1970)

593* ROBINSON, S.M.
COMPUTABLE ERROR BOUNDS FOR NONLINEAR PROGRAMMING
MRC TECHNICAL SUMMARY REPORT # 1274 UNIVERSITY OF
WISCONSIN, MADISON (1972)
PUBLIZIERT IN:
MATH. PROGRAMMING 5, 235-242 (1973)

594* ROBINSON, S.M.; NICKEL, K.
COMPUTATION OF THE PERRON ROOT AND VECTOR OF A
NON-NEGATIVE MATRIX
MRC TECHNICAL SUMMARY REPORT # 1100 UNIVERSITY OF
WISCONSIN, MADISON (1970)

ROBINSON, P.D.
SIEHE AUCH: BARNSLEY, M.F.; ROBINSON, P.D.

595* ROHN, J.
SYSTEMS OF LINEAR EQATIONS WITH COEFFICENTS
PRESCRIBED BY INTERVALS
EKONOM.-MAT. OBZOR. 12, 311-315 (1976)

596* ROHN, J.
INTERVALOVA ANALYZA LEONTEVOVA MODELU OBOR 11 - C5
- 9 AUTOREFERAT DISERTACE K ZISKANI VEDECKE
HODNOSTI KANDIDATA VED
MINISTERSTVO SKOLSTVI CSR VYSOKA SKOLA: UNIVERSITA
KARLOVA FAKULTA: MATEMATICKO-FYZIKALNI (1976)

597* ROHN, J.
INTERVALOVA ANALYZA LEONTEVOVA MODELU
KANDITATSKA PRACE, OBOR 11-05-9,
MATEMATICKO-FYZIKALNI FAKULTA KU, BEZ SKOLITELE
(1976)

598* ROHN, J.
CORRECTION OF COEFFICIENTS OF THE INPUT-OUTPUT
MODEL
GAMM-VORTRAG 1977, ERSCHEINT DEMNAECHST

599* ROKNE, J.
PRACTICAL AND THEORETICAL STUDIES IN NUMERICAL
ERROR ANALYSIS

PH. D. THESIS, DEP. OF MATH., STATIST. AND
COMPUTING SCIENCE, UNIVERSITY OF CALGARY (1969)

600* ROKNE, J.
FEHLERERFASSUNG BEI EIGENWERTPROBLEMEN VON
MATRIZEN
COMPUTING 7, 145-152 (1971)

601* ROKNE, J.
ERRORBOUNDS FOR SIMPLE ZEROS OF LAMBDA-MATRICES
RESEARCH PAPER 161, YELLOW SERIES DEP. OF MATH.,
STATIST. A. COMPUTING SCIENCE, THE UNIVERSITY OF
CALGARY (1972)

602* ROKNE, J.
EXPLICIT CALCULATION OF THE LAGRANGIAN INTERVAL
INTERPOLATING POLYNOMIAL
COMPUTING 9, 149-157 (1972)

603* ROKNE, J.
NEWTON'S METHOD UNDER MILD DIFFERENTIABILITY
CONDITIONS WITH ERROR ANALYSIS
NUMER. MATH. 18, 401-412 (1972)

604* ROKNE, J.
AUTOMATIC ERRORBOUNDS FOR SIMPLE ZEROS OF ANALYTIC
FUNCTIONS
COMM. ACM 16, 101-104 (1973)

605* ROKNE, J.
INTERVAL ARITHMETIC IN THE SOLUTION OF EQUATIONS
RESEARCH PAPER 217, YELLOW SERIES, DEP.OF MATH.,
STATIST. AND COMPUTING SCIENCE, UNIVERSITY OF
CALGARY (1974)

606* ROKNE, J.
REDUCING THE DEGREE OF AN INTERVAL POLYNOMIAL
COMPUTING 14, 5-14 (1975)

607* ROKNE, J.
ERRORBOUNDS FOR SIMPLE ZEROS OF LAMBDA-MATRICES
COMPUTING 16, 17-27 (1976)

ROKNE-SCHARF

608* ROKNE, J.
BOUNDS FOR AN INTERVAL POLYNOMIAL
COMPUTING 18, 225-240 (1977)

609* ROKNE, J.; LANCASTER, P.
AUTOMATIC ERRORBOUNDS FOR THE APPROXIMATE SOLUTION
OF EQUATIONS
COMPUTING 4, 294-303 (1969)

610* ROKNE, J.; LANCASTER, P.
COMPLEX INTERVAL ARITHMETIC
COMM. ACM 14, 111-112 (1971)

611* ROKNE, J.; LANCASTER, P.
ALGORITHM 86, COMPLEX INTERVAL ARITHMETIC
COMP. J. 18, 83-85 (1975)

612* ROOLF, K.
DIE AUFLOESUNG EINES LINEAREN GLEICHUNGSSYSTEMS
MIT TRIPLEX-ARITHMETIK
DIPLOMARBEIT AM INST. F. INFORMATIK, UNIVERSITAET
KARLSRUHE (1969)

ROTHMAIER, B.
SIEHE AUCH: BROCKHAUS, M.; ROTHMAIER, B.;
SCHROTH, P.

613* ROTHMAIER, B.
LANGZAHL-TRIPLEXARITHMETIK
DIPLOMARBEIT AM INST. F. INFORMATIK, UNIVERSITAET
KARLSRUHE (1968)

614* ROTHMAIER, B.
DOKUMENTATION DER STANDARDFUNKTIONEN DES
BETRIEBSSYSTEMS HYDRA X8
INTERNER BERICHT DES INST. F. INFORMATIK 70/8,
UNIVERSITAET KARLSRUHE (1970)

615* ROTHMAIER, B.
DOKUMENTATION UND BENUTZERANLEITUNG DER
LANGZAHLSTANDARDPROZEDUREN FUER DIE ELEKTRONISCHE
RECHENANLAGE EL-X8
INTERNER BERICHT DES INST. F. INFORMATIK 70/15,
UNIVERSITAET KARLSRUHE (1970)

616* ROTHMAIER, B.
DIE BERECHNUNG DER QUADRATWURZEL NEBST SCHRANKEN
AUF DUALMASCHINEN
INTERNER BERICHT DES INST. F. INFORMATIK 70/17,
UNIVERSITAET KARLSRUHE (1970)

617* ROTHMAIER, B.
DIE BERECHNUNG DER ELEMENTAREN FUNKTIONEN MIT
BELIEBIGER GENAUIGKEIT
DISSERTATION, INTERNER BERICHT DES INSTITUTS FUER
INFORMATIK 71/7, UNIVERSITAET KARLSRUHE (1971)

618* ROTHMAIER, B.
DER TRIPLEX-ALGOL COMPILER DER UNIVAC 1108
INTERNER BERICHT DES INST. F. PRAKT. MATH. 74/1,
UNIVERSITAET KARLSRUHE (1974)

619* SALENTIN, N.
LOESUNG LINEARER GLEICHUNGSSYSTEME HOHER ORDNUNG
MITTELS VERKETTETEM GAUSS-ALGORITHMUS IN
INTERVALLARITHMETIK
DIPLOMARBEIT AM INST. F. ANGEW. MATH. U.
INFORMATIK, UNIVERSITAET BONN (1970)

620* SCHARF, V.
EIN VERFAHREN ZUR LOESUNG DES CAUCHY-PROBLEMS FUER
LINEARE SYSTEME VON PARTIELLEN
DIFFERENTIALGLEICHUNGEN
DISSERTATION, UNIVERSITAET BONN (1967)

621* SCHARF, V.
UEBER EINE VERALLGEMEINERUNG DES
ANFANGSWERTPROBLEMS BEI LINEAREN SYSTEMEN VON
GEWOEHNLICHEN DIFFERENTIALGLEICHUNGEN
J. REINE ANGEW. MATH. 239/240, 287-290 (1970)

622* SCHARF, V.
UEBER EINE KLASSE LINEARER RANDWERTAUFGABEN
DIPLOMARBEIT BEI PROF. F. KRUECKEBERG, BONN

SCHEU-SCHMIDT

SCHEU, G.
SIEHE AUCH: ADAMS, E.; SCHEU, G.

623* SCHEU, G.
SCHRANKENKONSTRUKTION FUER DIE RANDWERTAUFGABE DER
LINEAREN WAERMELEITUNGSGLEICHUNG
DISSERTATION, UNIVERSITAET KARLSRUHE (1973)

624* SCHEU, G.
SCHRANKENKONSTRUKTION FUER DIE LOESUNG DER
ELLIPTISCHEN RANDWERTAUFGABE MIT KONSTANTEN
KOEFFIZIENTEN
Z. ANGEW. MATH. MECH. 55, T221-T223 (1975)

625* SCHEU, G.
ON THE OPTIMIZATION OF THE ERROR ESTIMATE OF A
FIXED POINT THEOREM
J. MATH. ANAL. APPL. 57, 298-322 (1977)

626* SCHEU, G.; ADAMS, E.
ZUR NUMERISCHEN KONSTRUKTION KONVERGENTER
SCHRANKEN-
FOLGEN FUER SYSTEME NICHTLINEARER, GEWOEHNLICHER
ANFANGSWERTAUFGABEN 'INTERVAL MATHEMATICS', ED. BY
K. NICKEL
LECTURE NOTES IN COMPUTER SCIENCE 29, SPRINGER
VERLAG, 279-287 (1975)

627* SCHEU, G.; ADAMS, E.
SCHRANKENKONSTRUKTION FUER DIE RANDWERTAUFGABE DER
LINEAREN, ELLIPTISCHEN DIFFERENTIALGLEICHUNG
DEUTSCHE LUFT- UND RAUMFAHRT FORSCHUNGSBERICHT
75-47 (1975)

628* SCHICHT, B.
VERFAHREN HOEHERER ORDNUNG ZUR EINSCHLIESSUNG DER
INVERSEN EINER KOMPLEXEN MATRIX
DIPLOMARBEIT AM INST. F. ANGEW. MATH.,
UNIVERSITAET KARLSRUHE (1974)

629* SCHILLING, U.
UEBER FIXPUNKTSAETZE IN VERALLGEMEINERTEN RAEUMEN
STAATSEXAMENSARBEIT AM INST. F. PRAKT. MATH.,
UNIVERSITAET KARLSRUHE (1974)

630* SCHIPKE, S.
EIN SPEZIELLES ROMBERG-VERFAHREN
WISS. Z. TECHN. UNIV. DRESDEN 25, 799-803 (1976)

631* SCHMIDT, J. W.
MONOTONE EINSCHLIESSUNG MIT DER REGULA FALSI BEI
KONVEXEN FUNKTIONEN
Z. ANGEW. MATH. MECH. 50, 640-643 (1970)

632* SCHMIDT, J. W.
EINGRENZUNG VON LOESUNGEN NICHTLINEARER
GLEICHUNGEN DURCH VERFAHREN MIT HOEHERER
KONVERGENZGESCHWINDIGKEIT
COMPUTING 8, 208-215 (1971)

633* SCHMIDT, J. W.
EINSCHLIESSUNG VON NULLSTELLEN BEI OPERATOREN MIT
MONOTON ZERLEGBARER STEIGUNG DURCH UEBERLINEAR
KONVERGENTE ITERATIONSVERFAHREN
ANN. ACAD. SCI. FENN. 502, 1-15 (1972)

634* SCHMIDT, J.W.
UEBERLINEAR KONVERGENTE MEHRSCHRITTVERFAHREN VOM
REGULA FALSI- UND NEWTON- TYP
Z. ANGEW. MATH. MECH. 53, 103-114 (1973)

635* SCHMIDT, J.W.
EIN ABLEITUNGSFREIES MEHRSCHRITTVERFAHREN FUER
EXTREMWERTAUFGABEN
MATH. NACHR. 59, 95-104 (1974)

636* SCHMIDT, J.W.
MONOTONE EINSCHLIESSUNG VON INVERSEN POSITIVER
ELEMENTE DURCH VERFAHREN VOM SCHULZ-TYP
COMPUTING 16, 211-219 (1976)

SCHMIDT-SCHUETH

637* SCHMIDT, J.W.; DRESSEL, H.
FEHLERABSCHAETZUNGEN BEI POLYNOMGLEICHUNGEN MIT
DEM FIXPUNKTSATZ VON BROUWER
NUMER. MATH. 10, 42-50 (1967)

638* SCHMIDT, J. W.; LEONHARDT, M.
EINGRENZUNG VON LOESUNGEN MIT HILFE DER REGULA
FALSI
COMPUTING 6, 318-329 (1970)

639* SCHMIDT, J.W.; NAUBER, W.
UEBER VERFAHREN ZUR ZWEISEITIGEN APPROXIMATION
INVERSER ELEMENTE
COMPUTING 17, 59-67 (1976)

640* SCHMITGEN, G.
TSCHEBYSCHEFF-APPROXIMATION VON
INTERVALLFUNKTIONEN DURCH VERALLGEMEINERTE
INTERVALLPOLYNOME
BERICHTE DER GMD BONN 26 (1970)

641* SCHMITGEN, G.
INTERVALLANALYTISCHE APPROXIMATIONSFRAGEN
Z. ANGEW. MATH. MECH. 51, T72-T73 (1971)

SCHMITT, B.
SIEHE: DUSSEL, R.; SCHMITT, B.

642* SCHMITT, B.
DIE KONVERGENZ NUMERISCHER ALGORITHMEN
DISSERTATION, UNIVERSITAET KARLSRUHE (1976)

643* SCHMITT, G.
SOME CONSIDERATIONS USING INTERVAL ANALYSIS IN
ADJUSTMENT COMPUTATIONS
XVI GENERAL ASSEMBLY OF THE INTERNAT. UNION OF
GEODESY AND GEOPHYSICS, INTERNAT. ASSOC. OF
GEODESY, GRENOBLE, AUGUST 1975 (1975)

SCHROEDER, G.
SIEHE AUCH: KRACHT, M.; SCHROEDER, G.
             RATSCHEK, H.; SCHROEDER, G.

644* SCHROEDER, G.
DIFFERENTIATIONSVERFAHREN FUER INTERVALLFUNKTIONEN
DISSERTATION, UNIVERSITAET DUESSELDORF (1971)

645* SCHROEDER, G.
DIFFERENTIATION OF INTERVAL FUNCTIONS
PROC. AMER. MATH. SOC. 36, 485-490 (1972)

646* SCHROEDER, G.
CHARAKTERISIERUNG DES QUASILINEAREN RAUMES I(R)
UND KLASSIFIZIERUNG DER QUASILINEAREN RAEUME DER
DIMENSION 1 UND 2
COMPUTING 10, 111-120 (1972)

647* SCHROEDER, G.
ZUM BEGRIFF DER DIMENSION VON QUASILINEAREN
RAEUMEN
WIRD VEROEFFENTLICHT

648* SCHROEDER, G.
ZUR STRUKTUR DES RAUMES DER HYPERNORMBAELLE UEBER
EINEM LINEAREN RAUM
COMPUTING 15, 67-70 (1975)

649* SCHROEDER, J.
UPPER AND LOWER BOUNDS FOR SOLUTIONS OF
GENERALIZED TWO-POINT BOUNDARY VALUE PROBLEMS
NUMER. MATH. 23, 433-457 (1975)

SCHROTH, P.
SIEHE: BROCKHAUS, M.; ROTHMAIER, B.; SCHROTH, P.

SCHUETH, H.
SIEHE: KANSY, K. ET AL.

SCHUHMACHER-SHIMIZU

SCHUHMACHER, H.
SIEHE AUCH: KANSY, K. ET AL.

650* SCHUHMACHER, H.
DER INTERVALLMATRIXBAUKASTEN
INTERNER BERICHT NR. 12, INST. F. NUMER.
DATENVERARBEITUNG, GMD BONN (1973)

651* SCHULDT, H.
MONOTONE VERFAHREN ZUR EINSCHLIESSUNG VON
NULLSTELLEN NICHTLINEARER GLEICHUNGEN
DIPLOMARBEIT AM INST. F. ANGEW. MATH. U.
INFORMATIK, UNIVERSITAET BONN (1970)

652* SCHULTHEISS, G.
LOESUNG LINEARER GLEICHUNGSSYSTEME AX=Y UNTER
GLEICHZEITIGER BERECHNUNG DER INVERSEN MATRIX MIT
EXAKTEN FEHLERSCHRANKEN
DIPLOMARBEIT AM INST. F. INFORMATIK, UNIVERSITAET
KARLSRUHE (1969)

653* SCHULZ, W.
POLYNOMAPPROXIMATIONEN UND APOSTERIORI
FEHLERSCHRANKEN FUER DIE LOESUNG DER
V. MISES-GRENZSCHICHT-GLEICHUNG
DIPLOMARBEIT AM INST. F. PRAKT. MATH.,
UNIVERSITAET KARLSRUHE (1973)

654* SCHWANENBERG, P.
ZUR NUMERISCHEN INTEGRATION VON GEWOEHNLICHEN
DIFFERENTIALGLEICHUNGEN MIT ANFANGSWERTMENGEN
DIPLOMARBEIT AM INST. F. ANGEW. MATH. U.
INFORMATIK, UNIVERSITAET BONN (1968)

655* SCHWANENBERG, P.
INTERVALLANALYTISCHE METHODEN ZUR LOESUNG VON
RANDWERTAUFGABEN BEI GEWOEHNLICHEN
DIFFERENTIALGLEICHUNGEN
BERICHTE DER GMD BONN 35 (1970)

656* SCHWANENBERG, P.
EIN NUMERISCHES VERFAHREN ZUR LOESUNG VON
RANDWERTAUFGABEN BEI GEWOEHNLICHEN
DIFFERENTIALGLEICHUNGEN
Z. ANGEW. MATH. MECH. 52, T220-221 (1972)

657* SCHWERMER, H.
ZUR FEHLERERFASSUNG BEI DER NUMERISCHEN
INTEGRATION VON GEWOEHNLICHEN
DIFFERENTIALGLEICHUNGSSYSTEMEN ERSTER ORDNUNG MIT
SPEZIELLEN ZWEIPUNKTVERFAHREN
'NUMERISCHE MATHEMATIK, DIFFERENTIALGLEICHUNGEN,
APPROXIMATIONSTHEORIE', ISNM 9, HERAUSG.
L.COLLATZ, G. MEINARDUS, H. UNGER, BIRKHAEUSER
VERLAG STUTTGART, BASEL, 141-155 (1968)

658* SCHWILL, W.-D.
FEHLERABSCHAETZUNG FUER DIE GEWOEHNLICHE
DIFFERENTIALGLEICHUNG 1.ORDNUNG UNTER
BERUECKSICHTIGUNG DER RUNDUNGSFEHLER
DISSERTATION, UNIVERSITAET KARLSRUHE (1972)

659* SHAMPINE, L.
MONOTONE ITERATIONS AND TWO-SIDED CONVERGENCE
SIAM J. NUMER. ANAL. 3, 607-615 (1966)

SHAYER, S.
SIEHE AUCH: MOORE, R.E.; DAVIDSON, J.A.;
JASCHKE, H.R.;
SHAYER, S.

660* SHAYER, S.
INTERVAL ARITHMETIC WITH SOME APPLICATIONS FOR
DIGITAL COMPUTERS
TECHNICAL REPORT 5-13-65-12
LOCKHEED MISSILES AND SPACE DIVISION, SUNNYVALE,
CAL. (1965)
AND:
M.A. THESIS, SAN JOSE STATE COLLEGE,
CAL. (1965)

661* SHIMIZU, T.
CONTRIBUTION TO THE THEORY OF NUMERICAL
INTEGRATION OF NON-LINEAR DIFFERENTIAL EQUATIONS
(III)
TRU MATHEMATICS 5, 51-66 (1969)

SHOKIN-SPREUER

662* SHOKIN, Y.
THE SECOND-ORDER INTERVAL-ANALYTIC METHOD FOR
ORDINARY DIFFERENTIAL EQUATIONS
VORTRAGSMANUSKRIPT, INTERNATIONALES SYMPOSIUM
UEBER INTERVALLMATHEMATIK, KARLSRUHE (1975)

663* SHOKIN, Y.; JULDASEV, Z.CH.
PREDSTAVIMOST' INTERVAL'NOSNACNYCH FUNKCIJ
VESCESTVENNYMI GRANICNYMI FUNKCIJAMI
CISLENNYE METODY MECHANIKI SPLOSNOJ SREDY 4,
134-146 (1973)

664* SHRIVER, B. D.
A SMALL GROUP OF RESEARCH PROJECTS IN MACHINE
DESIGN FOR SCIENTIFIC COMPUTATION
DAIMI-14, JUNE 1973, DEP. OF COMPUTER SCIENCE
UNIVERSITY OF AARHUS, DENMARK (1973)

665* SKELBOE, S.
COMPUTATION OF RATIONAL INTERVAL FUNCTIONS
BIT 14, 87-95 (1974)

666* SKELBOE, S.
TRUE WORST-CASE ANALYSIS OF LINEAR ELECTRICAL
CIRCUITS BY INTERVAL ARITHMETIC
INSTITUTE OF CIRCUIT THEORY AND TELECOMMUNICATION
TECHNICAL UNIVERSITY OF DENMARK, LYNGBY, REPORT IT
11 (1977)

667* SMITH, L. B.
INTERVAL ARITHMETIC DETERMINANT EVALUATION AND ITS
USE IN TESTING FOR A CHEBYSHEV SYSTEM
TECHNICAL REPORT CS 96, COMPUTER SCIENCE
DEPARTMENT, STANFORD UNIVERSITY (1968)
PUBLIZIERT IN:
COMM. ACM 12, 89-93 (1969)

SMITH, R.
SIEHE: HANSEN, E.; SMITH, R.

668* SOEHNER, H.-P.
DIE NUMERISCHE BEHANDLUNG VON REELLEN LINEAREN
INTERVALL-GLEICHUNGSSYSTEMEN MIT FEHLERERFASSUNG
FUER M-INTERVALLMATRIZEN
DIPLOMARBEIT AM INST. F. PRAKT. MATH. DER UNIV.
KARLSRUHE (1977)

669* SPANIOL, O.
DIE DISTRIBUTIVITAET IN DER INTERVALLARITHMETIK
COMPUTING 5, 6-16 (1970)

670* SPECK, P. T.
KREISARITHMETIK, REALISATION UND ANWENDUNG AUF
EINER RECHENANLAGE
DIPLOMARBEIT AM INST. F. ANGEW. MATH.,
UNIVERSITAET KARSLSRUHE (1973)

671* SPELLUCCI, P.
SIEHE AUCH: KRIER, N.; SPELLUCCI, P.

672* SPELLUCCI, P.; KRIER, N.
UNTERSUCHUNGEN DER GRENZGENAUIGKEIT VON
ALGORITHMEN ZUR AUFLOESUNG LINEARER
GLEICHUNGSSYSTEME MIT FEHLERERFASSUNG
'INTERVAL MATHEMATICS', ED. BY K. NICKEL
LECTURE NOTES IN COMPUTER SCIENCE 29, SPRINGER
VERLAG, 288-297 (1975)

673* SPELLUCCI, P.; KRIER, N.
EIN VERFAHREN ZUR BEHANDLUNG VON
AUSGLEICHSAUFGABEN MIT INTERVALLKOEFFIZIENTEN
COMPUTING 17, 207-218 (1976)

674* SPREKELS, J.
EXISTENZ UND ERSTE EINSCHLIESSUNG POSITIVER
LOESUNGEN BEI SUPERLINEAREN RANDWERTAUFGABEN
ZWEITER ORDNUNG
NUMER. MATH. 26, 421-428 (1976)

675* SPREUER, H.
SIEHE AUCH: ADAMS, E.; SPREUER, H.

SPREUER-STEWART

676* SPREUER, H.
EXISTENZ UND EINDEUTIGKEIT DER LOESUNG EINES
SYSTEMS VON INTEGRODIFFERENTIALGLEICHUNGEN FUER
DEN ENERGIETRANSPORT DURCH STRAHLUNG
DISSERTATION, UNIVERSITAET KARLSRUHE (1970)

677* SPREUER, H.
KONVERGENTE NUMERISCHE SCHRANKEN FUER PARTIELLE
RANDWERTAUFGABEN VON MONOTONER ART
'INTERVAL MATHEMATICS', ED. BY K. NICKEL
LECTURE NOTES IN COMPUTER SCIENCE 29, SPRINGER
VERLAG, 298-305 (1975)

678* SPREUER, H.; ADAMS, E.
SCHRANKEN FUER DIE LOESUNGEN PARABOLISCHER SYSTEME
Z. ANGEW. MATH. MECH. 51, T13-T14 (1971)

679* SPREUER, H.; ADAMS, E.
HINREICHENDE BEDINGUNG FUER AUSSCHLUSS VON
EIGENWERTEN IN PARAMETERINTERVALLE BEI EINER
KLASSE VON LINEAREN HOMOGENEN
FUNKTIONALGLEICHUNGEN
Z. ANGEW. MATH. MECH. 52, 479-485 (1972)

680* SPREUER, H.; ADAMS, E.
RANDWERTAUFGABEN VON MONOTONER ART MIT
RUECKKOPPLUNGS-UND NACHWIRKUNGSEFFEKTEN
Z. ANGEW. MATH. MECH. 52, T191-T193 (1972)

681* SPREUER, H.; ADAMS, E.; SRIVASTAVA, V. N.
MONOTONE SCHRANKENFOLGEN FUER GEWOEHNLICHE
RANDWERTAUFGABEN BEI SCHWACH GEKOPPELTEN
NICHTLINEAREN SYSTEMEN
Z. ANGEW. MATH. MECH. 55, 211-218 (1975)

682* SPRINKMEIER, I.
EINSCHLIESSUNG VON FUNKTIONEN DURCH
INTERVALLPOLYNOME MITTELS
TSCHEBYSCHEFF-APPROXIMATION
DIPLOMARBEIT AM INST. F. ANGEW. MATH. U.
INFORMATIK, UNIVERSITAET BONN (1970)

SRIVASTAVA, V. N.
SIEHE: SPREUER, H.; ADAMS, E.; SRIVASTAVA, V. N.

STATEMA, L.S.C.
SIEHE AUCH: EIJGENRAAM, P.; GRIEND, J.A.VAN DE;
STATEMA, L.S.C.

683* STEINLE, J.
KONVERTIERUNG VON GLEITKOMMAZAHLEN
DIPLOMARBEIT AM INST. F. ANGEW. MATH.,
UNIVERSITAET KARLSRUHE (1974)

STEWART, N. F.
SIEHE AUCH: DAVEY, D. P.; STEWART, N. F.

684* STEWART, N. F.
THE COMPARISON OF NUMERICAL METHODS FOR ORDINARY
DIFFERENTIAL EQUATIONS
TECHNICAL REPORT 3, DEPARTMENT OF COMPUTER
SCIENCE, UNIVERSITY OF TORONTO (1968)

685* STEWART, N. F.
GUARANTEED LOCAL ERROR BOUND FOR THE ADAMS METHOD
DEP. OF MATH. & STATISTICS, UNIVERSITY OF GUELPH,
ONTARIO (1970)

686* STEWART, N. F.
CERTAIN EQUIVALENT REQUIREMENTS OF APPROXIMATE
SOLUTIONS OF X' = F(T,X)
SIAM J. NUMER. ANAL. 7, 256-270 (1970)

687* STEWART, N. F.
CENTRALLY SYMMETRIC CONVEX POLYHEDRA
TO BOUND THE ERROR IN X' = F(T,X)
DOC. DE TRAVAIL PRES. AT THE SIAM 1971 NAT.
MEETING UNIVERSITY OF WASHINGTON, SEATTLE (1971)

688* STEWART, N. F.
A HEURISTIC TO REDUCE THE WRAPPING EFFECT
IN THE NUMERICAL SOLUTION OF X' = F(T,X)
BIT 11, 328-337 (1971)

689* STEWART, N. F.
CENTRALLY SYMMETRIC CONVEX POLYHEDRA TO BOUND THE
ERROR IN X'=F(T,X) (ABSTRACT)
SIAM REV. 14, 213 (1972)

690* STEWART, N. F.
INTERVAL ARITHMETIC FOR GUARANTEED BOUNDS IN
LINEAR PROGRAMMING
J. OF OPTIMAZATION THEORY AND APPLICATIONS 12, 1-5
(1973)

691* STEWART, N. F.
COMPUTABLE, GUARANTEED LOCAL ERROR BOUNDS FOR THE
ADAMS METHOD
MATH. NACHR. 60, 145-153 (1974)

692* STOUTEMYER, D.R.
AUTOMATIC ERROR ANALYSIS USING COMPUTER ALGEBRAIC
MANIPULATION
ACM TRANSACTIONS ON MATHEMATICAL SOFTWARE 3, 26-43
(1977)

693* STROEM, T.
STRICT ESTIMATION OF THE MAXIMUM OF A FUNCTION OF
ONE VARIABLE
BIT 11, 199-211 (1971)

694* STROEM, T.
AUTOMATISK KONSTRUKTION AV MAJORANTER MED
TILLAEMPNING PA FELUPPSKATTNINGAR VID APPROXIMATIV
BERAEKNING AV LINJAERA FUNKTIONALER
MANUSKRIPT

STROTHER, W.
SIEHE AUCH: MOORE, R.E.; STROTHER, W.; YANG, C.T.

695* STROTHER, W.
CONTINUITY FOR MULTI-VALUED FUNCTIONS AND SOME
APPLICATIONS TO TOPOLOGY
DOCTORAL DISSERTATION, TULANE UNIVERSITY (1952)

696* STROTHER, W.
FIXED POINT THEOREMS FOR MULTIVALUED FUNCTIONS
(NOTIZ)
BULL. AMER. MATH. SOC. 58, 60 (1952)

697* STROTHER, W.
FIXED POINTS, FIXED SETS, AND M-RETRACTS
DUKE MATH. J. 22, 551-556 (1955)

698* STROTHER, W.
CONTINUOUS MULTI-VALUED FUNCTIONS
BOL. SOC. MAT. SAO PAULO 10, 87-120 (1958)

699* STUMMEL, F.
STABILITY AND DISCRETE CONVERGENCE OF
DIFFERENTIABLE MAPPINGS
REV. ROUMAINE MATH. PURES APPL. 21, 63-96 (1976)

700* SUESSE, H.; OELSCHLAEGEL, D.
NULLSTELLENBESTIMMUNG MIT DER INTERVALLRECHNUNG
WISS. Z. TECHN. HOCHSCH. 'CARL SCHORLEMMER'
LEUNA-MERSEBURG 17, 423-432 (1975)

701* SUNAGA, T.
THEORY OF AN INTERVAL ALGEBRA AND ITS APPLICATION
TO NUMERICAL ANALYSIS
RAAG MEMOIRS 2, 29-46 (1958)

702* SUNAGA, T.
THEORY OF AN INTERVAL ALGEBRA
RAAG MEMOIRS 2, 547-564 (1958)

SURKAN, A.J.
SIEHE AUCH: FISCHER, H.; SURKAN, A.J.

703* TALBOT, T.D.
GUARANTEED ERROR BOUNDS FOR COMPUTED SOLUTIONS OF
NONLINEAR TWO-POINT BOUNDARY VALUE PROBLEMS
MRC TECHNICAL SUMMARY REPORT # 875 UNIVERSITY OF
WISCONSIN, MADISON (1968)

THIELER-ULLRICH

704* THIELER, P.
UNTERSUCHUNGEN ZU EINER
SENSITIVITAETS-INTERVALLARITHMETIK
DIPLOMARBEIT AM INST. F. ANGEW. MATH. U.
INFORMATIK, UNIVERSITAET BONN (1971)

705* THIELER, P.
SENSITIVITAET UND INVERTIERBARKEIT BEI MATRIZEN
DISSERTATION, UNIVERSITAET BONN (1974)

706* THIELER, P.
VERBESSERUNG VON FEHLERSCHRANKEN BEI ITERATIVER
MATRIZENINVERSION
'INTERVAL MATHEMATICS', ED. BY K. NICKEL
LECTURE NOTES IN COMPUTER SCIENCE 29, SPRINGER
VERLAG, 306-310 (1975)

707* THIELER, P.
EINE ANWENDUNG DES BROUWERSCHEN FIXPUNKTSATZES IN
DER INTERVALLARITHMETIK DER MATRIZEN
COMPUTING 14, 45-49 (1975)

TOFAHRN, W.
SIEHE AUCH: KANSY, K. ET AL.

708* TOFAHRN, W.
FEHLERERFASSUNG DURCH FORMALE ANWENDUNG VON
INTERVALLMETHODEN AM BEISPIEL DER
VAN-DER-POL-DIFFERENTIALGLEICHUNG
1. STAATSEXAMENSARBEIT, UNIVERSITAET BONN (1967)

709* TOST, R.
LOESUNG DER 1. RANDWERTAUFGABE DER
LAPLACE-GLEICHUNG IM RECHTECK MIT
INTERVALLANALYTISCHEN METHODEN
BERICHTE DER GMD BONN 28 (1970)

710* TOST, R.
INVERSION GROSSER MATRIZEN MIT INTERVALLMETHODEN
(NOTIZ)
Z. ANGEW. MATH. MECH. 50, T81 (1970)

711* TOST, R.
ZUR NUMERISCHEN LOESUNG VON RANDWERTAUFGABEN MIT
GESICHERTER FEHLEREINSCHLIESSUNG BEI PARTIELLEN
DIFFERENTIALGLEICHUNGEN
Z. ANGEW. MATH. MECH. 51, T74-175 (1971)

712* TREMEL, R.
MATHEMATIK MIT FEHLERN INTERVALL-RECHNUNG - EIN
NEUER ZWEIG DER MATHEMATIK ZAHLENINTERVALLE STATT
DISKRETER ZAHLEN
FRANKFURTER ALLGEM. ZEITUNG 28, NR. 152, SEITE 25
(14.7.1976)

713* TROTTER, W.T., JR.; BOGART, K.P.
MAXIMAL DIMENSIONAL PARTIALLY ORDERED SETS III: A
CHARACTERIZATION OF HIRAGUCHI'S INEQUALITY FOR
INTERVAL DIMENSION
DISCRETE MATHEMATICS 15, 389-400 (1976)

TURO, J.
SIEHE AUCH: KWAPISZ, M.; TURO, J.

714* TSAO, N. K.
ON THE DISTRIBUTION OF SIGNIFICANT DIGITS AND
ROUNDOFF ERRORS
COMM. ACM 17, 269-271 (1974)

715* ULLRICH, CH.
ALGORITHMEN ZUR INTERVALLMAESSIGEN BEHANDLUNG VON
LINEAREN GLEICHUNGSSYSTEMEN
DIPLOMARBEIT AM INST. F. ANGEW. MATH.,
UNIVERSITAET KARLSRUHE (1969)

716* ULLRICH, CH.
RUNDUNGSINVARIANTE STRUKTUREN MIT AEUSSEREN
VERKNUEPFUNGEN
DISSERTATION, UNIVERSITAET KARLSRUHE (1972)

717* ULLRICH, CH.
GESICHTSPUNKTE ZUR KOMPLEXEN RECHENARITHMETIK
Z. ANGEW. MATH. MECH. 55, T266-T268 (1975)

ULLRICH–WAUSCHKUHN

718* ULLRICH, CH.
UEBER DIE BEIM NUMERISCHEN RECHNEN MIT KOMPLEXEN
ZAHLEN UND INTERVALLEN VORLIEGENDEN MATHEMATISCHEN
STRUKTUREN
COMPUTING 14, 51-65 (1975)

719* ULLRICH, CH.
ZUM BEGRIFF DES RASTERS UND DER MINIMALEN RUNDUNG
COMPUTING SUPPLEMENTUM 1, ALBRECHT, R.;
KULISCH, U., SPRINGER-VERLAG WIEN, NEW YORK,
129-134 (1977)

720* ULLRICH, CH.
ZUR KONSTRUKTION KOMPLEXER KREISARITHMETIKEN
COMPUTING SUPPLEMENTUM 1, ALBRECHT, R.;
KULISCH, U., SPRINGER-VERLAG WIEN, NEW YORK,
135-150 (1977)

UNGER, H.
SIEHE: KRUECKEBERG, F.; UNGER, H.

721* URABE, M.
CONVERGENCE OF NUMERICAL ITERATION IN SOLUTION OF
EQUATIONS
J. SCI. HIROSHIMA UNIV. SER. A. 19, 479-489 (1956)

722* URABE, M.
ERROR ESTIMATION IN NUMERICAL SOLUTION OF
EQUATIONS BY ITERATION PROCESS
J. SCI. HIROSHIMA UNIV., SER. A-I 26, 77-91 (1962)

723* URABE, M.
COMPONENT-WISE ERROR ANALYSIS OF ITERATIVE METHODS
PRACTICED ON A FLOATING-POINT SYSTEM
MRC TECHNICAL SUMMARY REPORT #1268 UNIVERSITY OF
WISCONSIN, MADISON (1972)
PUBLIZIERT IN:
MEM. FAC. SCI. KYUSHU UNIV. SER. A, 27,
23-64 (1973)

724* URABE, M.
A POSTERIORI COMPONENT-WISE ERROR ESTIMATION OF
APPROXIMATE SOLUTIONS TO NONLINEAR EQUATIONS
'INTERVAL MATHEMATICS', ED. BY K. NICKEL
LECTURE NOTES IN COMPUTER SCIENCE 29, SPRINGER
VERLAG, 99-117 (1975)

725* URABE, M.
DEDICATION TO MINORU URABE
APPLICABLE ANAL. 6, 85-90 (1977)

VALENTINE, M.
SIEHE: GUDERLEY, K. G.; VALENTINE, M.

726* WALLACE, A. D.
CYCLIC INVARIANCE UNDER MULTI-VALUED MAPS
BULL. AMER. MATH. SOC. 55, 820-824 (1949)

727* WALLISCH, W.; GRUETZMANN, J.
INTERVALLARITHMETISCHE FEHLERANALYSE
BEITR. Z. NUMER. MATH. 3, 163-171 (1975)

728* WALSH, J.
ROUTINES FOR INTERVAL ARITHMETIC ON ATLAS
ATLAS COMPUTER LABORATORY, CHILTON, DIDCOT,
BERKSHIRE (1968)

729* WALZEL, A.
FEHLERABSCHAETZUNG BEI ANFANGSWERTAUFGABEN FUER
SYSTEME VON GEWOEHNLICHEN DIFFERENTIALGLEICHUNGEN
DISSERTATION, UNIVERSITAET KOELN (1969)

WARMBOLD, I.
SIEHE: NUDING, E.; WARMBOLD, I.;
KAHLERT-WARMBOLD, I.

730* WAUSCHKUHN, U.
METHODEN DER INTERVALLANALYSIS ZUR GLEICHMAESSIGEN
ERFASSUNG DES WERTEBEREICHS VON FUNKTIONEN IN
EINER UND MEHREREN VERAENDERLICHEN
DIPLOMARBEIT AM INST. F. ANGEW. MATH. U.
INFORMATIK, UNIVERSITAET BONN (1967)

WAUSCHKUHN-WISSKIRCHEN

731* WAUSCHKUHN, U.
PERIODISCHE LOESUNGEN BEI SYSTEMEN GEWOEHNLICHER
DIFFERENTIALGLEICHUNGEN
'3. INTERNAT. KONGRESS DATENVERARBEITUNG IM EUROP.
RAUM, BAND 2', HERAUSG. ARBEITSGEMEINSCHAFT
DATENVERARBEITUNG, NOVOGRAPHIC VERLAG, WIEN, 49-52
(1972)

732* WAUSCHKUHN, U.
INTERVALLANALYTISCHE METHODEN ZUM EXISTENZNACHWEIS
UND ZUR KONSTRUKTION DER LOESUNG VON
PERIODIZITAETSPROBLEMEN BEI GEWOEHNLICHEN
DIFFEPENTIALGLEICHUNGEN
BERICHTE DER GMD BONN 83 (1973)

733* WAUSCHKUHN, U.
BESTIMMUNG PERIODISCHER LOESUNGEN VON SYSTEMEN
GEWOEHNLICHER DIFFERENTIALGLEICHUNGEN MIT
INTERVALLANALYTISCHEN METHODEN
Z. ANGEW. MATH. MECH. 54, T237-T238 (1974)

WILHELM, J.
SIEHE: NUDING, E.; WILHELM, J.

734* WILHELM, J.
GRUNDOPERATIONEN UND IHRE BEHANDLUNG AUF
ELEKTRONISCHEN DATENVERARBEITUNGSANLAGEN
DIPLOMARBEIT AM INST. F. ANGEW. MATH.,
UNIVERSITAET HEIDELBERG (1976)

735* WILKER, H.-A.
LINEARE OPTIMIERUNG MIT FEHLERERFASSUNG DURCH
INTERVALLARITHMETIK
DIPLOMARBEIT, TECHN. UNIVERSITAET CLAUSTHAL (1974)

WILKINSON, J.H.
SIEHE: OETTLI, W.; PRAGER, W.; WILKINSON, J.H.

736* WIPPERMANN, H.-W.
MANUAL FUER DAS SYSTEM TRIPLEX-ALGOL KARLSRUHE
INSTITUT F. ANGEWANDTE MATHEMATIK - RECHENZENTRUM,
UNIVERSITAET KARLSRUHE (1967)

737* WIPPERMANN, H.-W.
REALISIERUNG EINER INTERVALL-ARITHMETIK IN EINEM
ALGOL60-SYSTEM
ELEKTRON. RECHENANLAGEN 9, 224-233 (1967)

738* WIPPERMANN, H.-W.
EIN ALGOL-60 COMPILER MIT TRIPLEX-ZAHLEN
Z. ANGEW. MATH. MECH. 47, T76-T79 (1967)

739* WIPPERMANN, H.-W.
DEFINITION VON SCHRANKENZAHLEN IN TRIPLEX-ALGOL
COMPUTING 3, 99-109 (1968)

740* WIPPERMANN, H.-W.
IMPLEMENTIERUNG EINES ALGOL-SYSTEMS MIT
SCHRANKENZAHLEN
ANGEW. INFORMATIK (ELEKTRON. DATENVERARBEITUNG)
10, 189-194 (1968)

741* WIPPERMANN, H.-W. (HERAUSGEBER)
APOSTOLATOS, N.; KRAWCZYK, R.; KULISCH, U.; LORTZ,
B.;
NICKEL, K.; WIPPERMANN, H.-W. THE ALGORITHMIC
LANGUAGE TRIPLEX-ALGOL-60
NUMER. MATH. 11, 175-180 (1968)

742* WISSKIRCHEN, P.
EIN STEUERUNGSPRINZIP DER INTERVALLRECHNUNG UND
DESSEN ANWENDUNG AUF DEN GAUSS'SCHEN ALGORITHMUS
BERICHTE DER GMD BONN 20 (1969)

743* WISSKIRCHEN, P.
BERUECKSICHTIGUNG VON MONOTONIEVERHAELTNISSEN BEI
DER INTERVALLARITHMETISCHEN BEHANDLUNG LINEARER
GLEICHUNGSSYSTEME
Z. ANGEW. MATH. MECH. 50, T79-T80 (1970)

744* WISSKIRCHEN, P.
ITERATIONSVERFAHREN BEI INKLUSIONSMONOTONEN
INTERVALLFUNKTIONEN
'INTERVAL MATHEMATICS', ED. BY K. NICKEL
LECTURE NOTES IN COMPUTER SCIENCE 29, SPRINGER
VERLAG, 311-315 (1975)

WISSKIRCHEN–ZEH

745* WISSKIRCHEN, P.
VERGLEICH INTERVALLARITHMETISCHER
ITERATIONSVERFAHREN
COMPUTING 14, 45-49 (1975)

WITT, T.H.
SIEHE: DARGEL, R.H.; LOSCALZO, F.R.; WITT, T.H.

746* WONGWISES, P.
EXPERIMENTELLE UNTERSUCHUNGEN ZUR NUMERISCHEN
AUFLOESUNG VON LINEAREN GLEICHUNGSSYSTEMEN MIT
FEHLERERFASSUNG
DISSERTATION, INTERNER BERICHT DES INST. F. PRAKT.
MATH. 75/1, UNIVERSITAET KARLSRUHE (1975)

747* WONGWISES, P.
EXPERIMENTELLE UNTERSUCHUNGEN ZUR NUMERISCHEN
AUFLOESUNG VON LINEAREN GLEICHUNGSSYSTEMEN MIT
FEHLERERFASSUNG
'INTERVAL MATHEMATICS', ED. BY K. NICKEL
LECTURE NOTES IN COMPUTER SCIENCE 29, SPRINGER
VERLAG, 316-325 (1975)

YANG, C.T.
SIEHE: MOORE, R.E.; STROTHER, W.; YANG, C.T.
        MOORE, R.E.; YANG, C.T.

YASUI, H.
SIEHE AUCH: OHNAKA, K.; YASUI, H.; KURO, T.

748* YASUI, T.
SIGNIFICANT DIGIT ARITHMETIC ON ILLIAC IV
ILLIAC IV REPORT 211 (1969)

YOHE, J.M.
SIEHE AUCH: LADNER, T.D.; YOHE, J.M.

749* YOHE, J.M.
BEST POSSIBLE FLOATING POINT ARITHMETIC
MRC TECHNICAL SUMMARY REPORT # 1054 UNIVERSITY OF
WISCONSIN, MADISON (1970)

750* YOHE, J. M.
RIGOROUS BOUNDS ON COMPUTED APPROXIMATIONS TO
SQUARE ROOTS AND CUBE ROOTS
MRC TECHNICAL SUMMARY REPORT # 1088 UNIVERSITY OF
WISCONSIN, MADISON (1970)

751* YOHE, J.M.
INTERVAL BOUNDS FOR SQUARE ROOTS AND CUBE ROOTS
COMPUTING 11, 51-57 (1973)

752* YOHE, J.M.
ROUNDINGS IN FLOATING-POINT ARITHMETIC
IEEE TRANSACTIONS ON COMPUTERS C-22, 577-586
(1973)

753* YOHE, J.M.
FOUNDATIONS OF FLOATING POINT COMPUTER ARITHMETIC
MRC TECHNICAL SUMMARY REPORT 1302
UNIVERSITY OF WISCONSIN, MADISON (1973)

754* YOHE, J.M.
SOFTWARE FOR INTERVAL ARITHMETIC: A REASONABLY
PORTABLE PACKAGE
MRC TECHNICAL SUMMARY REPORT # 1731 UNIVERSITY OF
WISCONSIN, MADISON (1977)

755* YOHE, J.M.
THE INTERVAL ARITHMETIC PACKAGE
MRC TECHNICAL SUMMARY REPORT # 1755 UNIVERSITY OF
WISCONSIN, MADISON (1977)

756* YOUNG, R. C.
THE ALGEBRA OF MANY-VALUED QUANTITIES
MATH. ANN. 104, 260-290 (1931)

757* ZEH, T.
EIN INTERVALLARITHMETISCHES VERFAHREN ZUR
BESTIMMUNG ALLER NULLSTELLEN EINER FUNKTION ZWEIER
VERAENDERLICHER IN EINEM RECHTECKIGEN GEBIET DES
R2
DIPLOMARBEIT, TECHNISCHE HOCHSCHULE DARMSTADT
(1972)

References

[1] E. ADAMS AND W. F. AMES, *On contracting interval iteration for nonlinear problems in* R^N, CAM7, Center for Applied Mathematics, University of Georgia, Athens, GA, 1978.

[1a] G. ALEFELD AND J. HERZBERGER, *Einführung in die Intervallrechnung*, K. H. Böhling, U. Kulisch, H. Maurer, eds., Bibliographisches Institut Mannheim-Wien-Zürich, 1974.

[2] G. ALEFELD, *Das symmetrische Einzelschrittverfahren bei linearen Gleichungen mit Intervallen als Koeffizienten*, Computing, 18 (1977), pp. 329–340.

[3] ——, *Über die Durchführbarkeit des Gaussschen Algorithms bei Gleichungen mit Intervallen als Koeffizienten*, Ibid., 1 Supp. (1977), pp. 15–19.

[3a] W. F. AMES AND M. GINSBERG, *Bilateral algorithms and their applications*, Computational Mechanics, J. T. Oden, ed., Springer-Verlag, New York, 1975, pp. 1–31.

[4] N. APOSTOLATOS, U. KULISCH, R. KRAWCZYK, B. LORTZ, K. NICKEL AND H.-W. WIPPERMANN, *The algorithmic language triplex-Algol 60*, Numer. Math., 11 (1968), pp. 175–180.

[5] N. APOSTOLATOS AND U. KULISCH, *Grundlagen einer Maschinenintervallarithmetik*, Computing, 2 (1967), pp. 89–104.

[6] W. APPELT, *Fehlereinschliessung für die Lösung einer Klasse elliptischer Randwertaufgaben*, Z. Angew. Math. Mech., 54 (1974), pp. 207–208.

[7] J. AVENHAUS, *Ein Verfahren zur einschliessung der Lösung des Anfangswert problems*, Computing, 8 (1971), pp. 182–190.

[8] K.-H. BACHMANN, *Untersuchungen zur Einschliessung der Lösungen von Systemen gewöhnlicher Differentialgleichungen*, Beitr. Numer. Math., 1 (1974), pp. 9–42.

[8a] F. BIERBAUM AND K.-P. SCHWIERZ, *A bibliography on interval mathematics*, J. Comput. Appl. Math., 4 (1978), no. 1.

[9] H. CHRIST, *Realisierung einer Maschinenintervallarithmetik auf beliebigen Algol 60 Compilern*, Elektron. Rechenanlagen, 10 (1968), pp. 217–222.

[10] W. CHUBA AND W. MILLER, *Quadratic convergence in interval arithmetic, Part I*, BIT, 12 (1972), pp. 284–290.

[11] J. W. DANIEL AND R. E. MOORE, *Computation and Theory in Ordinary Differential Equations*, W. H. Freeman, San Francisco, 1970.

[12] D. P. DAVEY AND N. F. STEWART, *Guaranteed error bounds for the initial value problem using polytope arithmetic*, Publication #109, Department d'Informatique, Université de Montréal, 1972.

181

[13] R. DUSSEL, *Einschliessung des Minimalpunktes einer streng Konvexen Function auf einem n-dimensionalen Quader*, Internal report no. 74, University Karlsruhe, Institute for Practical Mathematik, Karlsruhe, 1972.

[14] R. DUSSEL AND B. SCHMITT, *Die Berechnung von Schraken für den Wertebereich eines Polynoms in einem Intervall*, Computing, 6 (1970), pp. 35–60.

[15] K. ECKER AND H. RATSCHEK, *Intervallarithmetik mit Wahrscheinlichkeitsstruktur*, Angew. Informatik (Elektron. Datenverarbeitung), 14 (1972), pp. 313–320.

[15a] I. GARGANTINI, *Further applications of circular arithmetic: Schroeder-like algorithms with error bounds for finding zeros of polynomials*, SIAM J. Numer. Anal., 15 (1978), pp. 497–510.

[16] ———, *Parallel Laguerre iterations: The complex case*, Numer. Math., 26 (1976), pp. 317–323.

[17] I. GARGANTINI AND P. HENRICI, *Circular arithmetic and the determination of polynomial zeros*, Ibid., 18 (1972), pp. 305–320.

[18] D. I. GOOD AND R. L. LONDON, *Computer interval arithmetic: definition and proof of correct implementation*, J. Assoc. Comput. Mach., 17 (1970), pp. 603–612.

[19] J. H. GRAY AND L. B. RALL, *Newton: a general purpose program for solving nonlinear systems*, MRC Technical Summary Report #790, University of Wisconsin, Madison, 1967.

[20] ———, *INTE: A UNIVAC 1108/1110 program for numerical integration with rigorous error estimation*, MRC Technical Summary Report #1428, Mathematics Research Center, University of Wisconsin, Madison, 1975.

[21] ———, *Automatic Euler–Maclaurin integration*, Proceedings of the 1976 Army Numerical Analysis and Computers Conference, Army Research Office Report 76-3, 1976, pp. 431–444.

[21a] D. GREENSPAN, *Numerical approximation of periodic solutions of van der Pol's equation*, J. Math. Anal. Appl., 39 (1972), pp. 574–579.

[22] K. G. GUDERLEY AND C. L. KELLER, *A basic theorem in the computation of ellipsoidal error bounds*, Numer. Math., 19 (1972), pp. 218–229.

[23] E. HANSEN, *Interval arithmetic in matrix computations, Part I*, SIAM J. Numer. Anal., 2 (1965), pp. 308–320.

[24] ———, *On solving systems of equations using interval arithmetic*, Math. Comput., 22 (1968), pp. 374–384.

[25] ———, ED., *Topics in Interval Analysis*, Oxford University Press, London, 1969.

[26] ———, *On the solution of linear algebraic equations with interval coefficients*, Linear Algebra and Appl., 2 (1969), pp. 153–165.

[27] ———, *Interval forms of Newton's method*, Computing, 20 (1978), pp. 153–163.

[28] ———, *A globally convergent interval analytic method for computing and bounding real roots*, BIT, 18 (1978), pp. 415–424.

[29] ———, *Global optimization using interval analysis*, to appear.

[30] E. HANSEN AND R. SMITH, *Interval arithmetic in matrix computations, Part II*, SIAM J. Numer. Anal., 4 (1967), pp. 1–9.

[30a] G. W. HARRISON, *Compartmental models with uncertain flow rates*, Math. Biosci., 1979, to appear.

[31] M. HEGBEN, *Eine scaling-invariante Pivotsuche für Intervallmatrizen*, Computing, 12 (1974), pp. 99–106.

[32] ———, *Ein Iterationsverfahren, welches die optimale Intervalleinschliessung der Inversen eines M-matrixintervalls liefert*, Ibid., 12 (1974), pp. 107–115.

[33] P. HENRICI, *Discrete Variable Methods in Ordinary Differential Equations*, John Wiley, New York, 1962.

[34] S. HUNGER, *Intervallanalytische Defektabschätzung zur Lösung mit exakter Fehlcrerfassung bei Anfangswertaufgaben für Systeme gewöhnlicher Differentialgleichungen*, Z. Angew. Math. Mech., 52 (1972), pp. 208–209.

[35] F. HUSSAIN, *Zur Bedeutung der Intervallanalysis bei numerisch-geodätischen Rechnung*, AVN, 4 (1971), pp. 128–132.

[36] L. W. JACKSON, *Interval arithmetic error-bounding algorithms*, SIAM J. Numer. Anal., 12 (1975), pp. 223–238.

[37] U. JAHN, *Eine Theorie der Gleichungssysteme mit Intervall-Koeffizienten*, Z. Angew. Math. Mech., 54 (1974), pp. 405–412.

[38] ———, *Eine auf den Intervall-Zahlen fussende 3-wertige lineare Algebra*, Math. Nachr., 65 (1975), pp. 105–116.

[39] ———, *Intervall-wertigen Mengen*, Ibid., 68 (1975), pp. 115–132.

[40] W. H. KAHAN, *Circumscribing an ellipsoid about the intersection of two ellipsoids*, Canad. Math. Bull., 11 (1968), pp. 437–441.

[41] K. KANSY, *Ableitungsverträgliche Verallgemeinerung der Intervallpolynome*, GMD Nr. 70, Bonn, 1973.

[42] G. KEDEM, *Automatic differentiation of computer programs*, MRC Technical Summary Report #1697, Mathematics Research Center, University of Wisconsin, Madison, 1976.

[43] R. KRAWCZYK, *Newton-Algorithmen zur Bestimmung von Nullstellen mit Fehlerschranken*, Computing, 4 (1969), pp. 187–201.

[44] N. KRIER, *Komplexe Kreisarithmetik*, Z. Angew. Math. Mech., 54 (1974), pp. 225–226.

[45] F. KRÜCKEBERG, *Inversion von Matrizen mit Fehlererfassung*, Z. Angew. Math. Mech., 46 (1966), pp. T69–70.

[46] ———, *Defekterfassung bei gewöhnlichen und partiellen Differentialgleichungen*, ISNM 9, Birkhauser Verlag, 1968, pp. 69–82.

[47] D. KUBA AND L. B. RALL, *A UNIVAC 1108 program for obtaining rigorous error estimates for approximate solutions of systems of equations*, MRC Technical Summary Report #1168, University of Wisconsin, 1972.

[48] U. KULISCH, *Grundlagen des Numerischen Rechnens*, Reihe Informatik, 19, Bibliographisches Institut, Mannheim, 1976.

[48a] R. LOHNER AND E. ADAMS, *On initial value problems in R^N with intervals for both initial data and a parameter in the equations*, CAM 8, Center for Applied Mathematics, University of Georgia, Athens, GA, 1978.

[49] N. MACHOST, *Eine numerisch gesicherte Behandlung des Simplexverfahrens mit Intervallanalytischen Methoden*, Z. Angew. Math. Mech., 51 (1971), pp. 64–66.

[50] K. MADSEN, *On the solution of nonlinear equations in interval arithmetic*, BIT, 13 (1973), pp. 428–433.

[51] L. J. MANCINI, *Applications of interval arithmetic in signomial programming*, Tech. Report SOL 75-23, Dept. of Operations Research, Stanford University, Stanford, CA, 1975.

[52] L. J. MANCINI AND G. P. MCCORMICK, *Bounding global minima*, Math. of Operations Res., 1 (1976), pp. 50–53.

[53] U. MARCOWITZ, *Fehlerabschätzung bei Anfangsuretaugaben für Systeme von gewöhnlichen Differentialgleichunge mit Anwendung auf das REENTRY-Problem*, Numer. Math., 24 (1974), 249–275.

[54] W. MILLER, *On an interval arithmetic matrix method*, BIT, 12 (1972), pp. 213–219.

[55] ———, *Quadratic convergence in interval arithmetic, Part III*, Ibid., 12 (1972), pp. 291–298.

[56] R. E. MOORE, *Interval arithmetic and automatic error analysis in digital computing*, Ph.D. dissertation, Stanford University, October 1962; also Applied Math. and Statist. Lab. Report No. 25, Stanford University, Stanford, CA, 1962.

[57] ———, *Interval Analysis*, Prentice-Hall, Englewood Cliffs, NJ, 1966.

[58] ———, *Functional analysis for computers*, Funktionalanalytische Methoden der numerischen Mathematik, ISNM Vol. 12, Birkhauser Verlag, 1969, pp. 113–126.

[59] ———, *Mathematical Elements of Scientific Computing*, Holt, Rinehart and Winston, New York, 1975.

[60] ———, *On computing the range of values of a rational function of n variables over a bounded region*, Computing, 16 (1976), pp. 1–15.

[61] ———, *A test for existence of solutions to nonlinear systems*, SIAM J. Numer. Anal., 14 (1977), pp. 611–615.

[62] ———, *Bounding sets in function spaces with applications to nonlinear operator equations*, SIAM Rev., 20 (1978), pp. 492–512.

[63] ———, *A computational test for convergence of iterative methods for nonlinear systems*, SIAM J. Numer. Anal., 15 (1978), pp. 1194–1196.

[64] R. E. MOORE AND S. T. JONES, *Safe starting regions for iterative methods*, Ibid., 14 (1977), pp. 1051–1065.

[65] W. NEULAND, *Die intervallarithmetische Behandlung spezieller Fredholmscher Integralgleichungen*, Z. Angew. Math. Mech., 50 (1970), p. 69.

[65a] K. NICKEL, *Aufgaben von monotoner Art und Intervall-Mathermatik*, Z. Angew. Math. Mech., 57 (1977), pp. T294–295.

[65b] ———, *Bounds for the set of solutions of functional-differential equations*, MRC Technical Summary Report #1782, Mathematics Research Center, University of Wisconsin, Madison, 1977.

[65c] ———, *The lemma of Max Müller–Nagumo–Westphal for strongly coupled systems of parabolic functional-differential equations*, MRC Tech. Report #1800, Mathematics Research Center, University of Wisconsin, Madison, 1977.

[66] ———, *Quadratureverfahren mit Fehlerschranken*, Computing, 3 (1968), pp. 47–64.

[67] ———, *On the Newton method in interval analysis*, MRC Rept. #1136, Mathematics Research Center, University of Wisconsin, Madison, 1971.

[68] ———, *The contraction mapping fixed point theorem in interval analysis*, MRC Technical Summary Report #1334, University of Wisconsin, Madison, 1973.

[69] ———, *Stability and convergence of monotonic algorithms*, MRC Technical Summary Report #1340, Mathematics Research Center, University of Wisconsin, Madison, 1973.

[70] ———, ED., *Interval Mathematics*, Lecture Notes in Computer Science 29, Springer-Verlag, New York, 1975.

[71] K. NICKEL AND K. RITTER, *Termination criterion and numerical convergence*, SIAM J. Numer. Anal., 9 (1972), pp. 277–283.

[72] F. A. OLIVEIRA, *Interval analysis and two-point boundary value problems*, Ibid., 11 (1974), pp. 382–391.

[73] F. W. OLVER, *A new approach to error arithmetic*, Ibid., 15 (1978), pp. 368–393.

[74] J. M. ORTEGA AND W. C. RHEINBOLDT, *Iterative Solution of Nonlinear Equations in Several Variables*, Academic Press, New York, 1970.

[75] S. V. PARTER, *Solutions of a differential equation arising in chemical reactor processes*, SIAM J. Appl. Math., 26 (1974), pp. 687–716.

[76] L. B. RALL, ED., *Error in Digital Computation*, Vol. 1, John Wiley, New York, 1965. (Also Vol. 2, 1965.)

[77] ———, *Computational Solution of Nonlinear Operator Equations*, John Wiley, New York, 1969.

[77a] ———, *A comparison of the existence theorems of Kantorovich and Moore*, MRC TSR # 1944, University of Wisconsin, Madison, March 1979.

[78] H. RATSCHEK, *Über einige intervallarithmetische Grundbegriffe*, Computing, 4 (1969), pp. 43–55.

[79] ———, *Die Subdistributivität der Intervallarithmetik*, Z. Angew. Math. Mech., 51 (1971), pp. 189–192.

[80] ———, *Centered forms*, SIAM J. Numer. Anal., to appear.

[81] H. RATSCHEK AND G. SCHROEDER, *Über die Ableitung von intervallwertigen Funktionen*, Computing, 7 (1971), pp. 172–187.

[82] P. L. RICHMAN AND A. J. GOLDSTEIN, *A midpoint phenomenon*, J. Assoc. Comput. Mach., 20 (1973), pp. 301–304.

[83] P. D. ROBERTS AND A. BEN-ISRAEL, *A suboptimization method for interval linear programming: a new method for linear programming*, Linear Algebra and Appl., 3 (1970), pp. 383–405.

[84] S. M. ROBINSON, *Computable error bounds for nonlinear programming*, Math. Programming, 5 (1973), pp. 235–242.

[85] J. ROKNE, *Reducing the degree of an interval polynomial*, Computing, 14 (1975), pp. 5–14.

[86] G. SCHMITT, *Some considerations using internal analysis in adjustment computations*, XVI General Assembly of the International Union of Geodesy and Geophysics, Internat. Assoc. of Geodesy (Grenoble, August 1975).

[86a] J. SCHRÖDER, *Einschliessungsaussagen bei Differentialgleichungen*, Ansorge, Törnig, Ed., Numerische Lösungen nichtlinearer partieller Differential-und Integrodifferential-gleichungen, Springer-Verlag, Berlin, 1972, pp. 23–29.

[86b] ———, *Upper and lower bounds for solutions of generalized two-point boundary value problems*, Numer. Math., 23 (1978), pp. 433–457.

[87] G. SCHROEDER, *Differentiation of interval functions*, Proc. Amer. Math. Soc., 36 (1972), pp. 485–490.

[88] S. SKELBOE, *Computation of rational interval functions*, BIT, 14 (1974), pp. 87–95.

[89] ———, *True worst-case analysis of linear electrical circuits by interval arithmetic*, Report I T 11, Institute of Circuit Theory and Telecommunications, Technical University of Denmark, Lyngby, 1977.

[90] O. SPANIOL, *Die Distributivität in der Intervallarithmetik*, Computing, 5 (1970), pp. 6–16.

[91] N. F. STEWART, *A heuristic to reduce the wrapping effect in the numerical solution of* $x' = f(t, x)$, BIT, 11 (1971), pp. 328–337.

[92] ———, *Interval arithmetic for guaranteed bounds in linear programming*, J. Optimization Theory Appl., 12 (1973), pp. 1–5.

[93] ———, *Computable, guaranteed local error bounds for the Adams method*, Math. Nachr., 60 (1974), pp. 145–153.

[94] W. STROTHER, *Fixed points, fixed sets, and M-retracts*, Duke Math. J., 22 (1955), pp. 551–556.

[95] ———, *Continuous multi-valued functions*, Bol. Soc. Mat. Sao Paulo, 10 (1958), pp. 87–120.

[96] R. TOST, *Zur numerischen Lösung von Rendwertaufgaben mit gesicherten fehlereinsch-liessung bei partiellen Differentialgleichungen*, Z. Angew. Math. Mech., 51 (1971), pp. 74–75.

[97] J. VON NEUMANN AND H. GOLDSTINE, *Numerical inversion of matrices of high order*, Bull. Amer. Math. Soc., 53 (1947), pp. 1021–1099.

[98] U. WAUSCHKUHN, *Bestimmung periodischer Lösungen von Systemen gewöhnlicher Differentialgleichungen mit intervallanalytischen Methoden*, Z. Angew. Math. Mech., 54 (1974), pp. 237–238.

[99] P. WISSKIRCHEN, *Vergleich intervallarithmetischer Iterationsverfahren*, Computing, 14 (1975), pp. 45–49.

[100] J. M. YOHE, *The interval arithmetic package*, MRC Technical Summary Report #1755, Mathematics Research Center, University of Wisconsin, Madison, 1977.

Index

AUTHOR INDEX

SUBJECT INDEX